A Textbook of
Farm Machinery and Power Engineering

NIPA® GENX ELECTRONIC RESOURCES & SOLUTIONS P. LTD.

New Delhi-110 034

About the Authors

Er.Basavaraj, Kelappaji College of Agricultural Engineering and Technology, KAU, Tavanur, Malappuram District, Kerala. He has graduated B.Tech. (Agri.Engg.) from College of Agricultural Engineering, University of Agricultural Sciences, Raichur, Karnataka in 2013 and M.Tech.(Agri.Engg.) specialized in Farm Machinery and Power from Tamil Nadu Agricultural University(TNAU), Coimbatore, Tamilnadu in 2015. He was worked as Teaching Associate in College of Agricultural Engineering, PJTSAU, Sangareddy, Telangana from 2015-2017. Presently he is doing his Ph.D. studies from Kerala Agricultural University. He has qualified ICAR-JRF examination with 13th rank in 2013 and ICAR-SRF examination with 2nd rank in 2017. He been awarded as Young Scientist Award" in National Conference on Agricultural and Rural Innovations for Sustainable Empowerment (ARISE-2016) at Warangal, Telangana. He has guided two B.Tech. (Agricultural Engineering) theses. Published 13 full length research articles in national and international journals and published 14 abstracts and review articles in different conferences/symposiums.

Er. D. Srigiri Assistant Professor, College of Agricultural Engineering, ANGRAU, Madakasira, Ananthapuram District, Andhra Pradesh, He has graduated B.Tech. (Agri.Engg.) in College of Agricultural Engineering, Acharya N G Ranga Agricultural University in 2013 and M.Tech. (Agri.Engg.) specialized in Farm Machinery and Power from Acharya N G Ranga Agricultural University in 2015. He was worked as Teaching Associate in College of Agricultural Engineering, PJTSAU, Sangareddy, Telangana from 2015-2016. Presently he has been working as Assistant Professor in the Department of Farm Machinery and Power Enginnering, College of Agricultural Engineering, Madakasira. He has qualified ICAR-SRF examination with 5th rank in 2016. Published 6 full length research articles in national and international journals and published 5 abstracts in different conferences and 10 popular articles in various Telugu magazines. He has guided two B.Tech. (Agricultural Engineering) theses. He is also life member of Indian Society of Agricultural Engineers (ISAE) and Associate Member of Institute of Engineers India (IEI).

Dr. Jayan P.R, Professor (FPM), Department of Farm Machinery and Power Engineering, KCAET (KAU), Tavanur- 679 573. He has graduated in Agricultural Engineering from Kerala Agricultural University, Kelappaji College of Agricultural Engineering and Technology,Tavanur in 1990 and post graduate in Farm Power and Machinery in 1993 and Doctor of Philosophy in Farm Machinery and Powerfrom Tamilnadu Agricultural University, Coimbatore in 2003. He started his career as Assistant Engineer (Agri.) in the Department of Agriculture, Government of Kerala in 1993, later on appointed as Assistant Professor (Farm Power and Machinery) in Kelappaji College of Agricultural Engineering and Technology (KCAET), Kerala Agri. Univeristy (KAU)-Tavanur in 1994. He was the Head, Department of FPME, from 2009 to August, 2018 and presently continuing as Professor in the Department.

Dr. Jayan P. R received INAE - Award for his doctoral research work by the Indian National Academy of Engineering, New Delhi and appreciations for the efforts taken to develop various farm machineries under a State Plan project by Agri. Production Commissioner, Govt. of Kerala. He has been the Principal Investigator for many projects in the farm mechanization sector, member of the test code formulation for farm machines, Ministry of Agri. & Farmers Welfare, Department of AC & FW, M&T Division, Govt of India and member, editorial board, of IJAE, Muzaffar Nagar and Journal of Tropical Agriculture, KAU. He guided UG, PG, Ph.D. student projects as Major Advisor. He has published/ presented around 19 papers in international and national journals/ symposium/workshops and 23 popular articles in various magazines.

A Textbook of Farm Machinery and Power Engineering

Basavaraj
Kelappaji College of Agricultural Engineering and Technology
Kerala Agriculture University
Tavanur, Malappuram District, Kerala - 679573

D. Srigiri
Assistant Professor
College of Agricultural Engineering
ANGRAU, MADAKASIRA
Ananthapuram District, Andhra Pradesh - 515 301

Jayan P.R.
Professor (FPM)
Department of Farm Machinery and Power Engineering
Kelappaji College of Agricultural Engineering and Technology
Kerala Agriculture University
Tavanur, Malappuram District, Kerala - 679573

NIPA® GENX ELECTRONIC RESOURCES & SOLUTIONS P. LTD.
New Delhi-110 034

NIPA® GENX ELECTRONIC RESOURCES & SOLUTIONS P. LTD.

101,103, Vikas Surya Plaza, CU Block
L.S.C. Market, Pitam Pura, New Delhi-110 034
Ph : +91 11 27341616, 27341717, 27341718
E-mail: newindiapublishingagency@gmail.com
www: www.nipabooks.com

For customer assistance, please contact
Phone: + 91-11-27 34 17 17
Fax: + 91-11- 27 34 16 16
E-Mail: feedbacks@nipabooks.com

ISBN 978-81-960536-9-7

Composed & Designed by NIPA

Preface

This book has been written to meet the requirement of students getting knowledge in Agricultural Engineering and Farm Machinery and Power Engineering. A need has felt for a book on this subject of basic standards. A keen attempt has been made to make the book useful and interesting. Leaving all the unnecessary things only the essential facts and concepts have been discussed in this book.

It has a direct link with all the concepts in the farm machinery and power engineering that all the common and complicated concepts have been explained very short and clearly. It has been experienced that many students lack in their confidence to face examinations, therefore this book has been written in such way that it may equip the students with basic knowledge to face examinations.

This book is prepared by keeping the ARS-NET syllabus of Farm Power and Machinery discipline in mind and it contains excellent collection of important points on farm machinery, farm power, ergonomics, theory of machines, energy in agriculture, instrumentation and workshop technology tomeet requirements of students. The book should serve as a useful resource to the agricultural engineering and farm machinery and power engineering students appearing for various competitive exams such as ICAR JRF/SRF, NET,ARS and GATE etc. The book contains a section on key notes related to important terms on farm machinery and power engineering. It is useful for better understanding of this subject. The authors are deeply thankful to individuals, who helped on time for preparing this book.

It is hoped that this book will be useful for all the students of agricultural engineering as well as farm machinery and power engineering.

Any suggestions towards its further improvement will be thankfully acknowledged and incorporated in the next edition.

Basavaraj
D. Srigiri
Jayan P.R.

Contents

Syllabi for NET/ARS/ICAR/IARI Examinations of Farm Machinery and Power

1.1 ARS/NET Syllabus (FPM)

Unit 1: Farm Mechanization and Equipment

Status and scope of farm mechanization in India. Power availability on the farm. Identification of need based priorities of mechanization for various cropping systems. Hand tools used for different kinds of farm work, design considerations and materials for construction. Functional requirements, principles of working, construction, design, operation and management of animal-and power-operated equipment for tillage, land development, sowing, planting, fertilizer application, inter-cultivation, mowing, chaff cutting and baling. Special equipment for crops such as sugarcane, cotton, groundnut, and potato.

Unit 2: Design of Farm Machinery Components Design and selection of machinery elements: gear, pulleys, chains and sprockets, belts and simple clutches. Dynamic balancing and stability of farm machines. Force analysis on agricultural tools and implements. Pull, draft, unit draft and energy calculations for animal and power operated equipment. Machinery systems design.

Unit 3: Testing and Management of Farm Machinery Calibration of seed drills, planters, sprayers and fertilizer applicators. Performance and losses in harvesting and threshing. Calculations of field capacity, efficiency and rate of seed, fertilizer and chemicals applicators, threshers, harvesters and chaff cutters. Methods of testing of tillage equipment, seed-drills, seeders, planters, sprayers, threshers, and combines. Farm machinery selection for different soils, crops and operations. Cost analysis of implements and operations. Estimation of power-energy requirements. Reliability of farm machinery.

Unit 4: Engines and Tractor Systems Engineering thermodynamics. Various systems of spark and compression ignition engines. Operations, adjustment and troubleshooting on the working of the systems. Calculations on horse power, torque, speed, firing arrangement and intervals, heat load and power transmission from piston to the fly wheel. Tractor transmission, Types of clutch, gear trains, differential and final drives. Tractor chassis mechanics. Tractor power outlets. Mechanical and power steering systems. Hydraulics and hitching systems, ADDC. Tractor performance tests. Maintenance schedules of tractors and power tillers. Recent trends in tractor design.

Unit 5: Ergonomics and Safety Anthropometry in equipment design, physiological cost and effect of work on physiological responses, fatigue and comfort; ergonomics in design of farm tools; safety aspects of agricultural machinery; effect of noise and vibration on work performance; chemical hazards and control measures; operator's protective gadgets; 125 design of tractor controls viz., hand and foot controls, visual range and limitations, seat design, etc.

Unit 6: Soil Dynamics in Tillage and Traction Dynamic properties of soil and their measurements; stress-strain relationships; theories of soil failure, mechanics of tillage tools; design parameters and performance of tillage tools. Introduction to traction devices, tyre function and size, their selection, mechanics of traction devices, traction theories, slippage and sinkage of wheels, evaluation and prediction of traction performance; soil compaction - causes and methods for alleviating the effect on soil and crop responses.

Unit 7: Manufacturing Technology Specification of materials, surface roughness, production drawing, computer aided drawing heat treatment, workshop practices applied in prototype production, common tools and press operations, metal cutting and machining, jigs, fixtures and gauges, casting and die-casting processes; basic joining processes, welding processes, testing of joints and metallurgy.

Unit 8: Instrumentation and Measurement Techniques Mechanical measurements, sensors and transducers, application of electrical strain gauges, signal transmission and processing, dynamic measurements; measurement of temperature, pressure, strain, force, torque, power vibrations etc.; determination of calorific value, fluid flow rates etc; signal conditioning and monitoring, data acquisition and storage.

Unit 9: Energy in Agriculture Conventional and renewable energy sources in agriculture; solar radiation and its measurement; characteristics of solar spectrum; solar energy collection, storage and applications; solar photovoltaic conversion and SPV powered systems. Types of wind mills and their applications; thermo-chemical conversion of biomass, direct combustion, Pyrolysis and gasification, chemical conversion processes, carbonization, briquetting, pelletization and densification of biomass; bioconversion into alcohols, methyl and ethyl esters, organic acids, solvents of amino acids; types of biogas plants, biogas properties, uses and distribution, alternate fuels for IC engines. Energy requirement in agricultural production systems, energy ratio and specific energy value, inflow and outflow of energy in unit agricultural operation, energy audit, accounting and analysis.

1.2 ICAR SRF

Agricultural Engineering and Technology

1.2.1 Farm Machinery and Power

Unit 1: Farm Mechanization and Equipment Status and scope of farm mechanization in India. Power availability on the farm. Identification of need based priorities of mechanization for various cropping systems. Hand tools used for different kinds of farm work, design considerations and materials for construction. Functional requirements, principles of working, construction, design, operation and management of animal-and power-operated equipment for tillage, land development, sowing, planting, fertilizer application, inter-cultivation, mowing, chaff cutting and baling. Special equipment for crops such as sugarcane, cotton, groundnut, and potato.

Unit 2: Design of Farm Machinery Components Design and selection of machinery elements: gear, pulleys, chains and sprockets, belts and simple clutches. Dynamic balancing and stability of farm machines. Force analysis on agricultural tools and implements. Pull, draft, unit draft and energy calculations for animal and power operated equipment. Machinery systems design.

Unit 3: Testing and Management of Farm Machinery Calibration of seed drills, planters, sprayers and fertilizer applicators. Performance and losses in harvesting and threshing. Calculations of field capacity, efficiency and rate of seed, fertilizer and chemicals applicators, threshers, harvesters and chaff cutters. Methods of testing of tillage equipment, seed-drills, seeders, planters, sprayers, threshers, and combines. Farm machinery selection for different soils, crops and operations. Cost analysis of implements and operations. Estimation of power-energy requirements. Reliability of farm machinery.

Unit 4: Engines and Tractor Systems Engineering thermodynamics. Various systems of spark and compression ignition engines. Operations, adjustment and trouble shooting on the working of the systems. Calculations on horse power, torque, speed, firing arrangement and intervals, heat load and power transmission from piston to the fly wheel. Tractor transmission, Types of clutch, gear trains, differential and final drives. Tractor chassis mechanics. Tractor power outlets. Mechanical and power steering systems. Hydraulics and hitching systems,

ADDC. Tractor performance tests. Maintenance schedules of tractors and power tillers. Recent trends in tractor design.

Unit 5: Ergonomics and Safety Anthropometry in equipment design, physiological cost and effect of work on physiological responses, fatigue and comfort; ergonomics in design of farm tools; safety aspects of agricultural machinery; effect of noise and vibration on work performance; chemical hazards and control measures; operator's protective gadgets; design of tractor controls viz., hand and foot controls, visual range and limitations, seat design, etc.

Unit 6: Soil Dynamics in Tillage and Traction Dynamic properties of soil and their measurements; stress-strain relationships; theories of soil failure, mechanics of tillage tools; design parameters and performance of tillage tools. Introduction to traction devices, tyre function and size, their selection, mechanics of traction devices, traction theories, slippage and sinkage of wheels, evaluation and prediction of traction performance; soil compaction - causes and methods for alleviating the effect on soil and crop responses.

Unit 7: Manufacturing Technology Specification of materials, surface roughness, production drawing, computer aided drawing heat treatment, workshop practices applied in prototype production, common tools and press operations, metal cutting and machining, jigs, fixtures and gauges, casting and die-casting processes; basic joining processes, welding processes, testing of joints and metallurgy.

Unit 8: Instrumentation and Measurement Techniques Mechanical measurements, sensors and transducers, application of electrical strain gauges, signal transmission and processing, dynamic measurements; measurement of temperature, pressure, strain, force, torque, power vibrations etc.; determination of calorific value, fluid flow rates etc; signal conditioning and monitoring, data acquisition and storage.

Unit 9: Energy in Agriculture Conventional and renewable energy sources in agriculture; solar radiation and its measurement; characteristics of solar spectrum; solar energy collection, storage and applications; solar photovoltaic conversion and SPV powered systems. Types of wind mills and their applications; thermo-chemical conversion of biomass, direct combustion, Pyrolysis and gasification, chemical conversion processes, carbonization, briquetting, pelletization and densification of biomass; bioconversion into alcohols, methyl and ethyl esters, organic acids, solvents of amino 82 acids; types of biogas plants, biogas properties, uses and distribution, alternate fuels for IC engines. Energy requirement in agricultural production systems, energy ratio and specific energy value, inflow and outflow of energy in unit agricultural operation, energy audit, accounting and analysis.

1.3 IARI Ph.D Entrance Examination

Agricultural Engineering

1.3.1 General

Basic concepts in calculus, trigonometry, analytical geometry, linear algebra and algebra of real and complex numbers; instrumentation for measurement of forces, torque, temperature, moisture, fluid flow; basic principles of simulation; methods of statics, dynamics and mechanics of materials; common distributions of random variables and methods of statistical inference; energy sources - their utilization and efficiencies on the farm; uses and application of computers. In addition, attempt any one of the following three areas depending upon the major field of choice.

1.3.2 Farm Power and Equipment

Farm Power and Equipment State of farm mechanization; testing of power units and tractor systems; performance capacities of power and machines on the farm; management of power and machinery and their use on the farm; dynamics of machine elements; tillage and tractor machines; design considerations in farm machinery and power units; tractor hydraulics, symbols and circuits; ergonomic consideration in machine design.

1

Introduction of Farm Power Machinery and Energy

Agricultural mechanization involves the design, manufacture, distribution, use and servicing of all types of agricultural tools, equipment and machines. It includes three main power sources: human, animal and mechanical with special emphasis on mechanical (tractive power).

Mechanization has been identified as one of the critical inputs for production agriculture. It may be described as an appropriate package of technology to i) ensure timely field operations to increase productivity, reduce crop losses and improve product quality, ii) increase land utilization and input use efficiency and iii) increase labour productivity through labour saving and drudgery reducing mechanical devices. The mechanization of Indian agriculture has proceeded along two-pronged approach based on improved equipment and enhanced power supply. However, compared to the mechanization of western agriculture which was motivated by the need to substitute human labour and draught animals with mechanical prime movers, the guiding principle in mechanizing Indian agriculture has been to maintain a socially desirable mix of human labour, draught animal power and mechanical power.

Scope of Mechanization

It is quite true that the farmers of developing countries have the lowest earnings per capita because of the low yield per hectare they get from their land holdings. One of the few important means of increasing farm production per hectare is to mechanize it. Mechanization may have to be done at various levels. Broadly, it can be done in three different ways:

- By introducing the improved agricultural implements on small size holdings to be operated by bullocks.
- By using the small tractors, tractor-drawn machines and power tillers on medium holdings to supplement existing sources.
- By using the large size tractors and machines on the remaining holdings to supplement animal power source.

The progress of the country should be mainly judged on the basis of degree of farm mechanization (production per worker and the horsepower under his command per unit area). Large amount of labour or draft power, which can be replaced through machines, provides a strong incentive to mechanize.

Sources of Farm Power and Mechanization

It is quite true that the Indian farmers have the lowest income per capita because of the low yield per hectare they get from their land holdings. One of the few important reasons of increasing farm production per hectare is to mechanize it. Mechanization in India may have to be done at various levels. Broadly, it can be done in three different ways: By introducing the improved agricultural implements on small size holdings to be operated by bullocks. By using the small tractors, tractor-drawn machines and power tillers on medium holdings to supplement existing sources. By using the large size tractors and machines on the remaining holdings to supplement animal power source.

Various types of agricultural operations performed on a farm can be broadly classified as: i) Tractive work such as seed bed preparation, cultivation, harvesting and transportation, and ii) Stationary work like silage cutting, feed grinding, threshing, winnowing and lifting of irrigation water.

These operations are performed by different sources of power namely, human, animal, stationary engine, tractor, power tiller, electricity, solar and wind.

For doing these operations different types of power available is classified as:

- Human power
- Animal power
- Mechanical power
- Electrical power
- Wind power

Different animal sources are: Bullocks- can pull of about 15% of its weight, Buffaloes, Camels, Horses and Donkeys-can pull 80 % of its weight for short period and 10-15% of its weight for sustainable period. The average force a bullock can exert is nearly equal to one tenth of its body weight. But for a very short period, it can exert many more times the average force. Generally a medium size bullock can develop between 0.50 to 0.75 hp.

2

Materials of Construction

Classification of Materials

Broadly there are two groups of materials used for making farm machines namely metallic and non-metallic:

Metallic		Non-Metallic
Ferrous	Non-Ferrous	Wood (timber)
Iron : Cast iron		
Wrought iron	Copper (Cu)	Rubber
Steel: Steel		
Steel Alloys	Zinc (Zn)	Leather
	Lead (Pb)	Plastics
	Tin (Sn)	Fiberglass
	Aluminum (Al)	Teflon
	Brass (Cu+Zn)	
	Bronze (Cu+tin)	
	White metal or	
	Babbit (Tin+Cu)	
	Solder (Tin+lead)	

Carbon Content

Low Carbon (upto 0.25% C)

(Easy to bend, forge and shear)

Medium Carbon Steel (0.25% to 0.50% C)

High carbon steel (0.50 to 1.2% C)

Purpose of Alloying Substances

- Manganese: Used for cutting tools and high grade ball bearings. It imparts hardness.

- Chromium: Same as manganese.
- Tungsten: Helps in maintaining sharp edge even when the cutting tool is hot.
- Nickel: Helps in making steel elastic and ductile.

Application of Steel

- Low carbon steel: Used extensively in the construction of farm machinery. Frames and most of other members are made out of low-carbon steel.
- Medium carbon steel: Used for greater strength and hardness. Members such as shafts, connecting rods etc. are made of medium carbon steel.
- High carbon steel: It is very hard and is used for making tools, ball and roller bearings, cutting tools *etc.*

Non-Ferrous Metals: uses of non-ferrous metals and alloys are discussed below

- Copper (Cu): tubing, wires, windings *etc.*
- Brass: It is an alloy of Cu (60-90%) and Zn (10-40%). Used for making radiator pipes, brass welding rods, screen for fuel lines, instrument parts etc.
- Bronze: It is an alloy of Cu (80-95%) and Tin (5-20%) Zinc is also added sometimes. Used for making bushings, springs, pipe fittings, valves, pump pistons and bearings.
- Babbit: it is a tin-based alloy with small amount of copper Cu (7-8%), Antimony (8-9%) and Tin (80-84%). Used for making bearings.

3

Tillage Methods

It is a mechanical manipulation of soil to provide favourable condition for crop production. Soil tillage consists of breaking the compact surface of earth to a certain depth and to loosen the soil mass, so as to enable the roots of the crops to penetrate and spread into the soil.

Tillage operations following primary tillage which are performed to create proper soil tilth for seeding and planting arc sccondary tillage. These are lighter and finer operations, performed on the soil after primary tillage operations. Secondary tillage consists of conditioning the soil to meet the different tillage objectives of the farm. These operations consume less power per unit area compared to primary tillage operations. Secondary tillage implements may be tractor drawn or bullock drawn implements.

- Primary tillage: The operations performed to open up any cultivable land with a view to prepare seed bed for growing crops
- Primary tillage is done with country plough, mould board plough, disc plough, chisel plough, sub soiler plough and rotary plough
- Lighter and finer operations performed on the soil after primary tillage is called as secondary tillage operations
- The secondary tillage includes breaking the soil clods and levelling the land.
- Cultivators, harrows, sweeps, clod crushers and levellers are some of the secondary tillage implements
- The stubble mulch tillage is adopted to reduce wind and water erosion and to conserve water by reducing run off.
- The size of bullock drawn mould board ploughs is in the order of 150 mm and that of tractor drawn ploughs ranges from 300 to 400 mm.
- Throat clearance is the vertical distance between the share point and the beam of the plough.

- Depth of ploughing by disc plough can be increased by
  - Increasing disc angle
  - Decreasing tilt angle
  - Adding dead weight
- Furrow is the trench formed by a plough in the soil.
- Furrow slice is the soil mass cut, lifted and thrown to the side.
- Furrow wall is the unploughed vertical face of the furrow.
- Back furrow is a raised ridge left at the center of field in the round and round method of ploughing when ploughing is done from centre towards the sides.
- Dead furrow is an open trench left in between two adjacent strips of land after ploughing is completed when ploughing is done from sides towards the centre.
- Headland is the strip of field left unploughed along the edges of fields to facilitate turning of farm equipment.
- When a plough works round a strip of ploughed land, it is said to be gathering.
- When a plough works round a strip of unploughed land, it is said to be casting.
- Pull is the force required to operate a farm equipment.
- Draft is the horizontal component of pull in the direction of travel.
- Draft is expressed in kgf or N.
- Side draft is the horizontal component of pull perpendicular to the direction of travel.
- For better pulverization with Mb plough, stubble bottom type mould board is used
- Chisel plough can be used for 30-45cm deep ploughing
- Subsoilers are designed to operate to a depth of about 90 cm
- The specific energy requirements of rotary tiller is increased by increasing bite length
- Coulters are used with mould board plough

- Indigenous plough is a multi-purpose tool
- A cultivator whose tynes back when obstacles encounter is called spring tooth cultivator
- A bullock drawn implement have lower working capacity at low speed
- The concave end of disc harrow tends to penetrate more than the convex side
- The forces acting upon side and bottom of land side as well as rolling resistance of supporting wheels of a plough are known as parasitic force
- The part of the plough that penetrate into the soil and cut the soil in horizontal direction is called share
- The front edge of the share which makes horizontal cut is cutting edge
- The forward end of cutting edge that actually penetrate into soil and cut the soil is called share point
- In bar point share, the bar serves as landside
- The share of MB plough is made of chilled cast iron
- A type of share provided with adjustable bar is bar point share
- The MB plough works on the principle of suction
- The stubble type mould board is suitable for grassy lands
- A long mould board with gentle curvature used in tough soils full of grass is called general purpose mould board
- The slat type mould board is used in sticky soils
- Tilt angle of standard disc plough is 15-25°
- The tilt angle of wheat plough is 0°
- The type of furrow opener recommended for use in hard or trashy ground and also in wet sticky soil is single disc type furrow opener
- The part of MB plough to which all other parts of plough bottom attached is called frog
- In MB plough a small irregular piece of metal has similar shape to ordinary plough bottom is jointer
- A part of MB plough which lifts, turns and break furrow slice is mould board

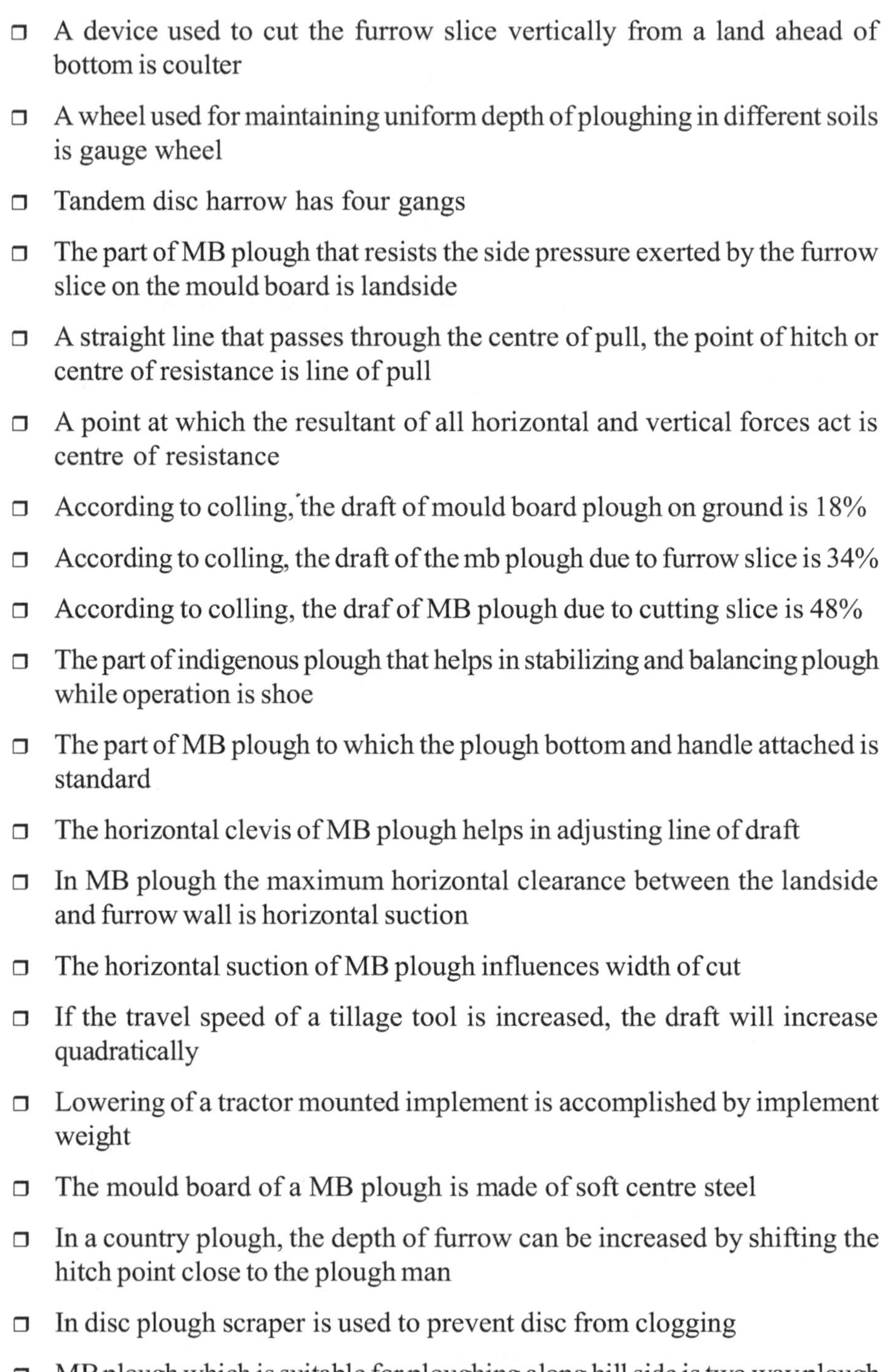

- A device used to cut the furrow slice vertically from a land ahead of bottom is coulter
- A wheel used for maintaining uniform depth of ploughing in different soils is gauge wheel
- Tandem disc harrow has four gangs
- The part of MB plough that resists the side pressure exerted by the furrow slice on the mould board is landside
- A straight line that passes through the centre of pull, the point of hitch or centre of resistance is line of pull
- A point at which the resultant of all horizontal and vertical forces act is centre of resistance
- According to colling, the draft of mould board plough on ground is 18%
- According to colling, the draft of the mb plough due to furrow slice is 34%
- According to colling, the draf of MB plough due to cutting slice is 48%
- The part of indigenous plough that helps in stabilizing and balancing plough while operation is shoe
- The part of MB plough to which the plough bottom and handle attached is standard
- The horizontal clevis of MB plough helps in adjusting line of draft
- In MB plough the maximum horizontal clearance between the landside and furrow wall is horizontal suction
- The horizontal suction of MB plough influences width of cut
- If the travel speed of a tillage tool is increased, the draft will increase quadratically
- Lowering of a tractor mounted implement is accomplished by implement weight
- The mould board of a MB plough is made of soft centre steel
- In a country plough, the depth of furrow can be increased by shifting the hitch point close to the plough man
- In disc plough scraper is used to prevent disc from clogging
- MB plough which is suitable for ploughing along hill side is two way plough

- The depth measured at the centre of the disc by placing its concave side down on a flat surface is called concavity
- In disc harrow spool is used to provide distance between two discs mounted on same gang axle
- A shaft of disc harrow on which the number of discs are mounted is called arbor bolt
- The increase in tilt angle of disc influences penetration of disc
- The type of bearing generally used to support the discs on a standard disc plough is tapered roller bearing
- A MB plough that throws slice to one side of direction of travel is called one way plough
- A MB plough that throws slice to both sides of direction of travel is called two way plough
- A MB plough whose plough bottom turns through 180° about a longitudinal axis is turn wrest plough
- In a standard disc plough, all the discs are mounted on separate standards
- The type of restrained three point linkage system in which the depth of implement is automatically adjusted to maintain a pre-selected constant draft is called automatic draft control system
- The method of ploughing in which the plough works round the strip of ploughed land is gathering
- The method of ploughing in which the plough works round the strip of unploughed land is casting
- An open trench left in between two adjacent strips of land after ploughing is called dead furrow
- A disc harrow with four gangs in which a set of two gangs follow behind the set of other two gangs is tandem disc harrow
- A harrow which is based on the principle that side thrust against the front gang is opposed by side thrust of rear gang is offset disc harrow
- A raised ridge left at centre of strip of land when ploughing is started from centre to side is back furrow
- The internal angle between the gangs of single action disc harrow is more than 90°

- The internal angle between the gangs of double action disc harrow is less than 90°
- The offset disc harrow are mostly suitable for working in orchards
- The size of disc harrow is determined by maximum width of cut
- A bullock drawn harrows mostly have rigid tynes
- A ridger is used for making furrows
- Heavy draft of a disc harrow is due to furrows too wide
- Animal drawn patella is clod crusher
- Sweeps are used for mulching
- Normal ploughing depth is 15 cm
- A subsoiler is used to break a hard pan of soil upto a depth of 100 cm
- The shape of furrow cut by a indigenous lough is V shape
- The shape of furrow cut by MB plough is L shape
- The indigenous plough is used for Ploughing, harvesting tuber crops and puddling
- No till system reduces runoff by 90%
- The condition of soil that determines the optimum condition for tillage is crumbly
- Land plane is ideally suited for top finishing
- A relatively short/broad mould board that is curved rather abruptly near the top resulting in a greater degree of pulverization is called stubble bottom type mould board
- Surface planting with conservation feature of contour lifting is called ridge planting
- The hitching of plough is done by placing the plough few centimetres below the ground level
- The useful life of walking type plough is about 1500 h
- The tillage practice sub soiling is desirable in the areas of hard or compacted layers below the plough depth
- The versatile implement used for seed bed preparation, ridge making and channel shaping is hoe

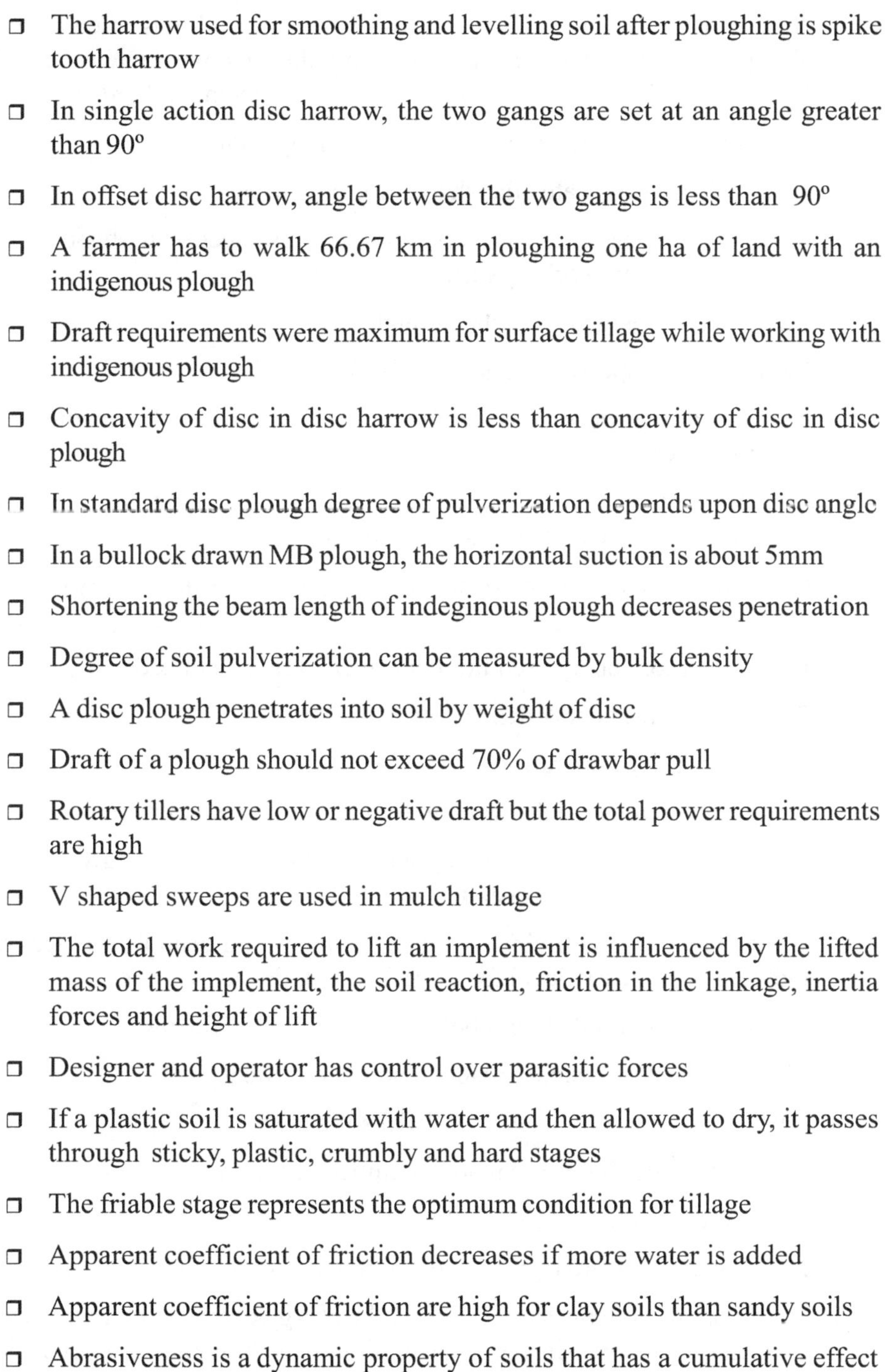

- The harrow used for smoothing and levelling soil after ploughing is spike tooth harrow
- In single action disc harrow, the two gangs are set at an angle greater than 90°
- In offset disc harrow, angle between the two gangs is less than 90°
- A farmer has to walk 66.67 km in ploughing one ha of land with an indigenous plough
- Draft requirements were maximum for surface tillage while working with indigenous plough
- Concavity of disc in disc harrow is less than concavity of disc in disc plough
- In standard disc plough degree of pulverization depends upon disc angle
- In a bullock drawn MB plough, the horizontal suction is about 5mm
- Shortening the beam length of indeginous plough decreases penetration
- Degree of soil pulverization can be measured by bulk density
- A disc plough penetrates into soil by weight of disc
- Draft of a plough should not exceed 70% of drawbar pull
- Rotary tillers have low or negative draft but the total power requirements are high
- V shaped sweeps are used in mulch tillage
- The total work required to lift an implement is influenced by the lifted mass of the implement, the soil reaction, friction in the linkage, inertia forces and height of lift
- Designer and operator has control over parasitic forces
- If a plastic soil is saturated with water and then allowed to dry, it passes through sticky, plastic, crumbly and hard stages
- The friable stage represents the optimum condition for tillage
- Apparent coefficient of friction decreases if more water is added
- Apparent coefficient of friction are high for clay soils than sandy soils
- Abrasiveness is a dynamic property of soils that has a cumulative effect rather than an immediate effect

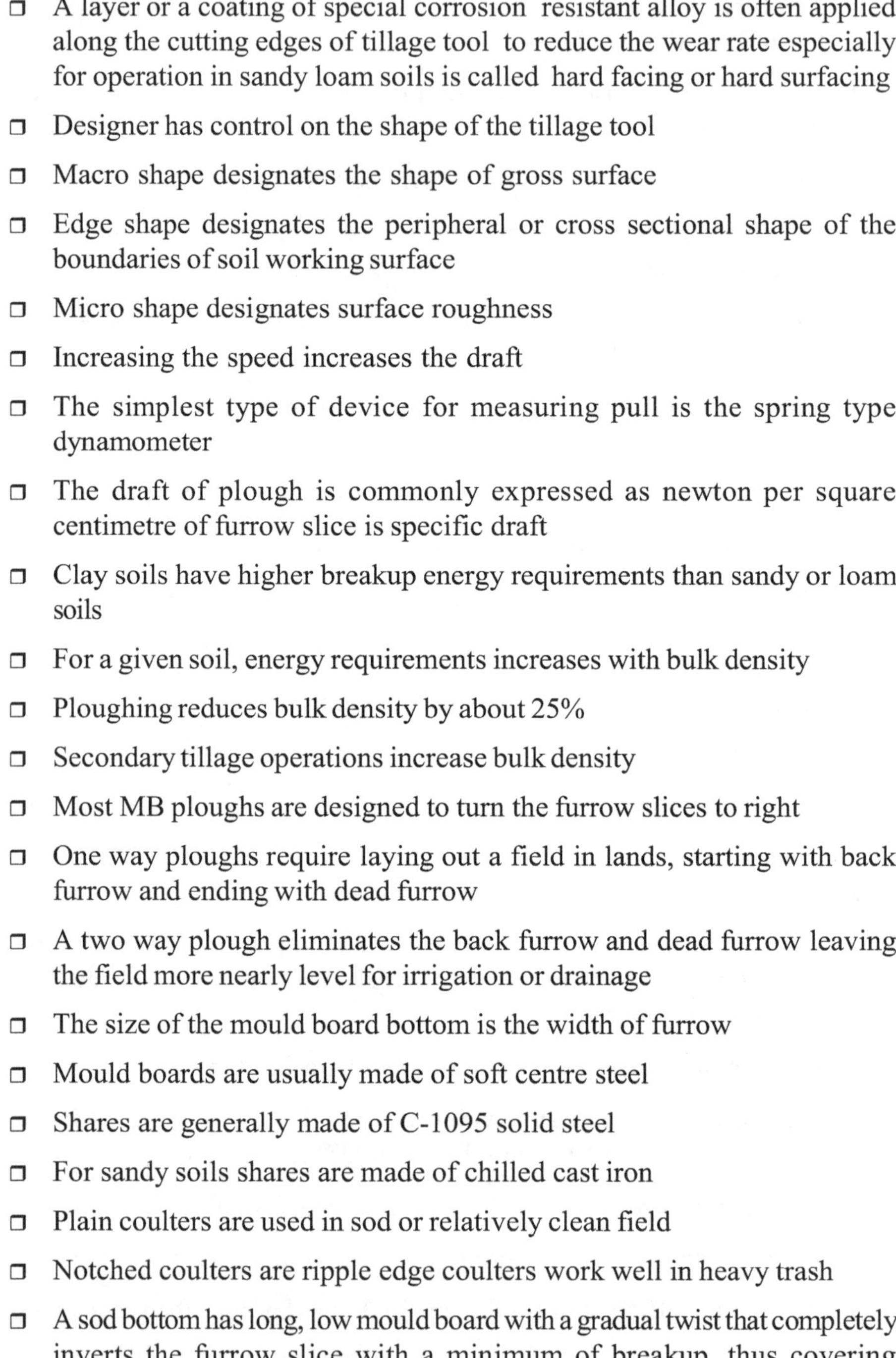

- A layer or a coating of special corrosion resistant alloy is often applied along the cutting edges of tillage tool to reduce the wear rate especially for operation in sandy loam soils is called hard facing or hard surfacing
- Designer has control on the shape of the tillage tool
- Macro shape designates the shape of gross surface
- Edge shape designates the peripheral or cross sectional shape of the boundaries of soil working surface
- Micro shape designates surface roughness
- Increasing the speed increases the draft
- The simplest type of device for measuring pull is the spring type dynamometer
- The draft of plough is commonly expressed as newton per square centimetre of furrow slice is specific draft
- Clay soils have higher breakup energy requirements than sandy or loam soils
- For a given soil, energy requirements increases with bulk density
- Ploughing reduces bulk density by about 25%
- Secondary tillage operations increase bulk density
- Most MB ploughs are designed to turn the furrow slices to right
- One way ploughs require laying out a field in lands, starting with back furrow and ending with dead furrow
- A two way plough eliminates the back furrow and dead furrow leaving the field more nearly level for irrigation or drainage
- The size of the mould board bottom is the width of furrow
- Mould boards are usually made of soft centre steel
- Shares are generally made of C-1095 solid steel
- For sandy soils shares are made of chilled cast iron
- Plain coulters are used in sod or relatively clean field
- Notched coulters are ripple edge coulters work well in heavy trash
- A sod bottom has long, low mould board with a gradual twist that completely inverts the furrow slice with a minimum of breakup, thus covering vegetative matter thoroughly

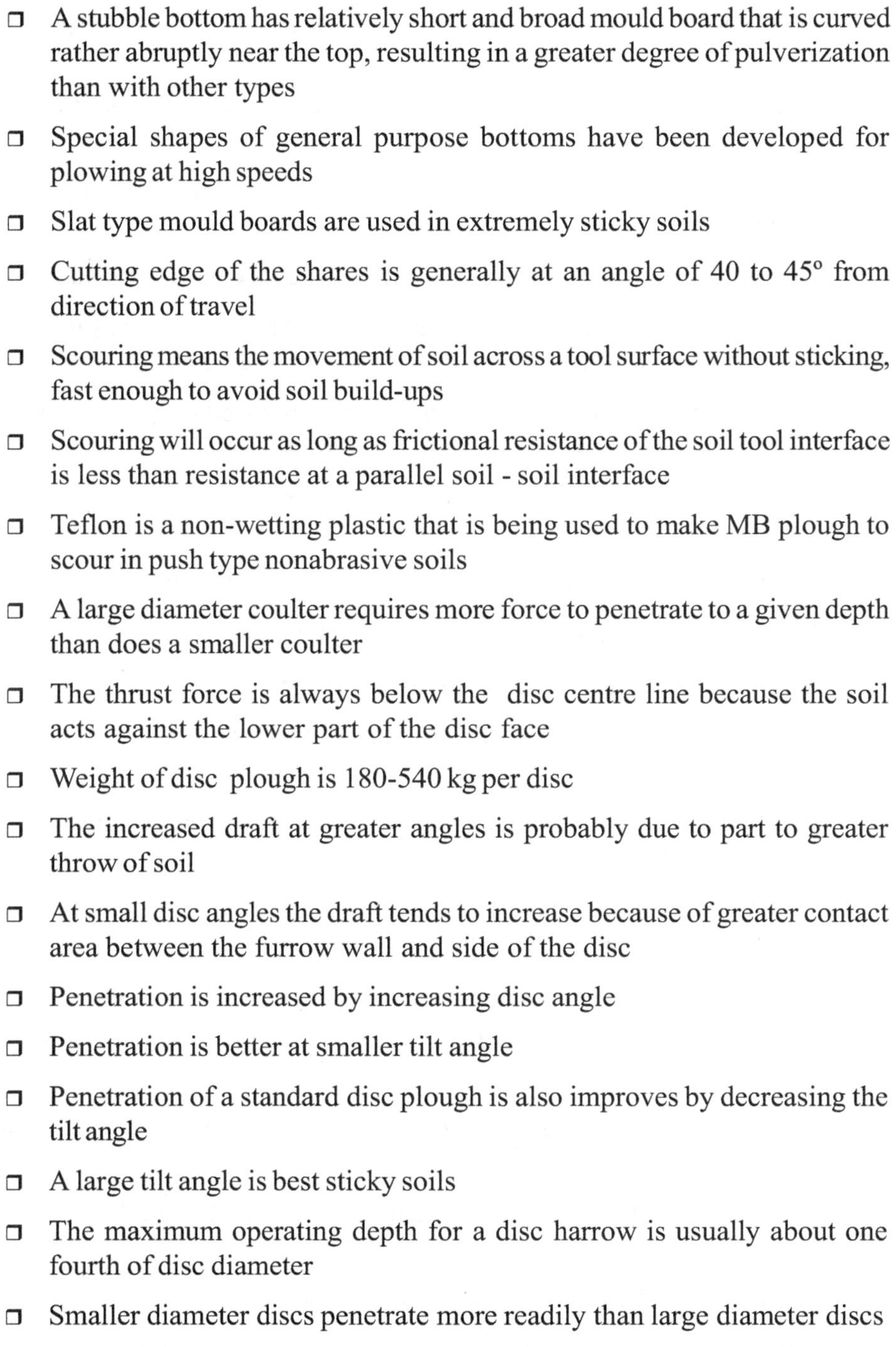

- A stubble bottom has relatively short and broad mould board that is curved rather abruptly near the top, resulting in a greater degree of pulverization than with other types
- Special shapes of general purpose bottoms have been developed for plowing at high speeds
- Slat type mould boards are used in extremely sticky soils
- Cutting edge of the shares is generally at an angle of 40 to 45° from direction of travel
- Scouring means the movement of soil across a tool surface without sticking, fast enough to avoid soil build-ups
- Scouring will occur as long as frictional resistance of the soil tool interface is less than resistance at a parallel soil - soil interface
- Teflon is a non-wetting plastic that is being used to make MB plough to scour in push type nonabrasive soils
- A large diameter coulter requires more force to penetrate to a given depth than does a smaller coulter
- The thrust force is always below the disc centre line because the soil acts against the lower part of the disc face
- Weight of disc plough is 180-540 kg per disc
- The increased draft at greater angles is probably due to part to greater throw of soil
- At small disc angles the draft tends to increase because of greater contact area between the furrow wall and side of the disc
- Penetration is increased by increasing disc angle
- Penetration is better at smaller tilt angle
- Penetration of a standard disc plough is also improves by decreasing the tilt angle
- A large tilt angle is best sticky soils
- The maximum operating depth for a disc harrow is usually about one fourth of disc diameter
- Smaller diameter discs penetrate more readily than large diameter discs
- Cut out blades penetrate a little better than plain blades because of reduced peripheral contact area

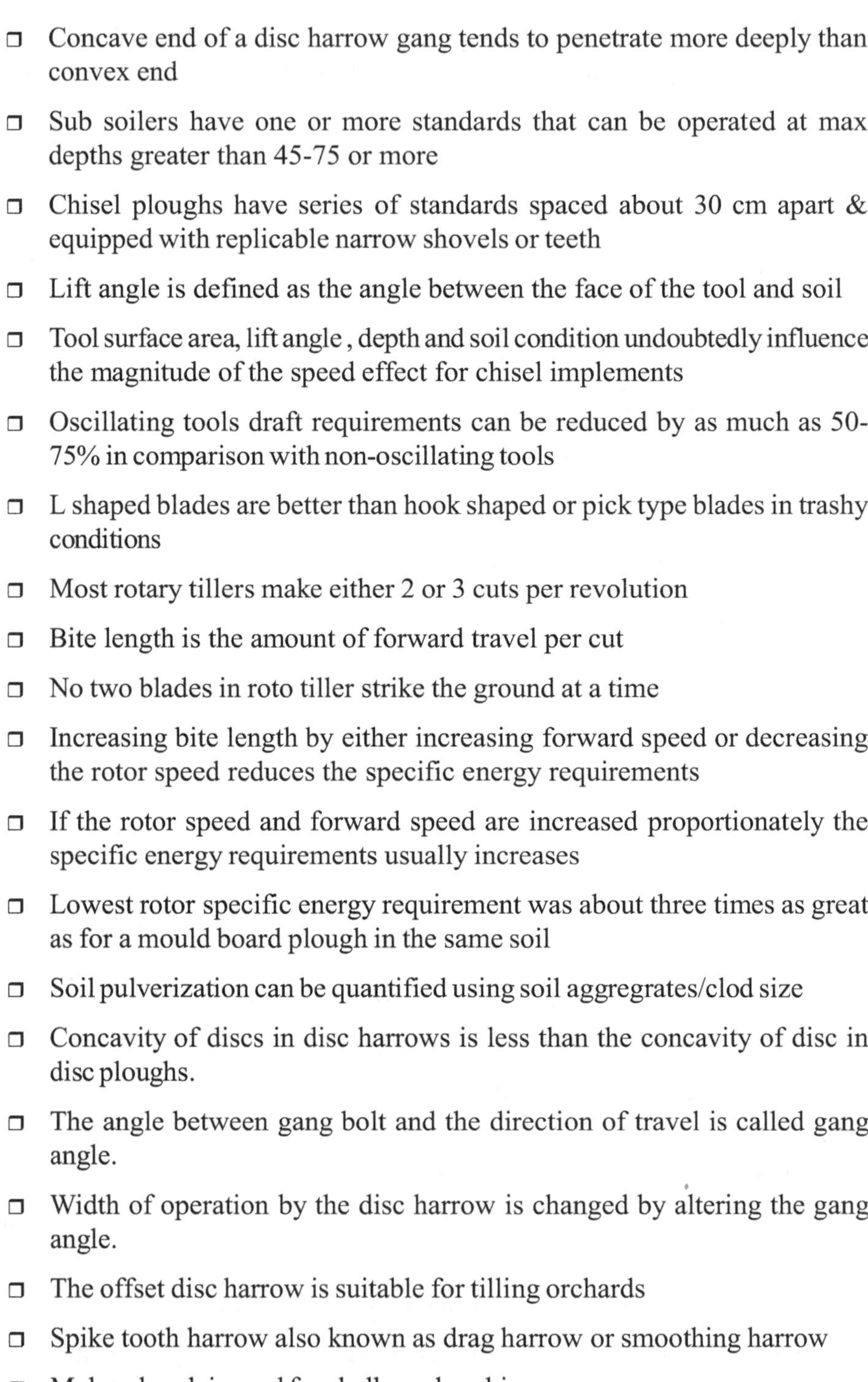

- Concave end of a disc harrow gang tends to penetrate more deeply than convex end
- Sub soilers have one or more standards that can be operated at max depths greater than 45-75 or more
- Chisel ploughs have series of standards spaced about 30 cm apart & equipped with replicable narrow shovels or teeth
- Lift angle is defined as the angle between the face of the tool and soil
- Tool surface area, lift angle , depth and soil condition undoubtedly influence the magnitude of the speed effect for chisel implements
- Oscillating tools draft requirements can be reduced by as much as 50-75% in comparison with non-oscillating tools
- L shaped blades are better than hook shaped or pick type blades in trashy conditions
- Most rotary tillers make either 2 or 3 cuts per revolution
- Bite length is the amount of forward travel per cut
- No two blades in roto tiller strike the ground at a time
- Increasing bite length by either increasing forward speed or decreasing the rotor speed reduces the specific energy requirements
- If the rotor speed and forward speed are increased proportionately the specific energy requirements usually increases
- Lowest rotor specific energy requirement was about three times as great as for a mould board plough in the same soil
- Soil pulverization can be quantified using soil aggregrates/clod size
- Concavity of discs in disc harrows is less than the concavity of disc in disc ploughs.
- The angle between gang bolt and the direction of travel is called gang angle.
- Width of operation by the disc harrow is changed by altering the gang angle.
- The offset disc harrow is suitable for tilling orchards
- Spike tooth harrow also known as drag harrow or smoothing harrow
- Melur plough is used for shallow ploughing

- Sheep foot roller is used in wet lands after puddling to arrest deep percolation losses
- The performance index of a farm equipment is directly proportional to the area covered per unit time
- A power tiller is most suited for rotary cultivation because its traction requirement is less, generates negative draft and provides higher degree of soil pulverization
- Puddling is done to reduce percolation of water
- L shaped blades are best suited for minimum tillage of trashy lands at shallow depths
- Centre of load of a 40 X 40 cm size MB plough from wing of share of its first plough bottom is at 90cm
- A harrow that has one right hand gang and one left hand gang operating in tandem is called offset disc harrow
- Depth of working of a tillage implement is controlled by gauge wheel
- Power tillers operate most satisfactorily with rotovator
- Material used for frog is cast iron
- Degree of soil pulverization by MB plough can be measured by mean volume diameter of soil particles
- Ridge shaped furrow bed is obtained using ridger and ridge height should be less than 15% od depth of cut
- Optimum speed of rotor in rotavator is 150-350 rpm
- Soil pulverization can be quantified using soil aggregrates/ clod size
- Cone index reflects the strength of soil
- Cone index is a composite parameter
- Cone penetration is measured at a constant penetration rate of 33mm/s
- The draft & total power requirement of rotary cultivator operating in concurrent mode as compared to a spring tyne cultivator of equal cutting width under some operating conditions respectively lower & higher.
- Soil compaction in track type tractor in comparison to wheel type tractor of same weight is less

- Soil is more susceptible to compaction when its moisture content is approximately at dry level
- A farmer can justify owing a particular machine by estimating the approximate break-even point owing a machine and hiring a custom operator.
- Utility index is a measure of work machine contact hours
- The labour cost per hectare is inversely proportional to the field capacity of the machine.

4

Sowing and Planting

Seeding or sowing is an art of placing seeds in the soil to have good germination in the field. A perfect seeding gives a) Correct amount of seed per unit area. b) Correct depth at which seed is placed in the soil. c) Correct spacing between row-to-row and plant-to-plant. Seed drill is a machine for placing the seeds in a continuous flow in furrows at uniform rate and at controlled depth with or without the arrangement of covering them with soil. Planter is normally used for those seeds which are larger in size and cannot be used by usual seed drills. A planter consists of: (i) hopper (ii) feed metering device (iii) knock out arrangement (iv) cut-off mechanism (v) furrow opener and (vi) other accessories. A planter has seed hopper for each row.

- The equipment used for dropping seeds in a continuous stream and the spacing between plant to plant is not constant is seed drill.
- The method of planting in which row to row as well as plant to plant distance is uniform is called check row planting.
- Broadcasting is random scattering of seeds on the surface of the field.
- Hill dropping is placing group of seeds at about equal interval in rows.
- Furrow planting or lister planting is practiced under semi-arid conditions for row crops.
- Bed planting is suitable for high rainfall areas.
- Flat planting is generally predominated where natural moisture conditions are favourable.
- The horizontal plate type planter is the most common example of cell type planter.
- Edge cell, edge drop plates are well suited for planting large seeds like corn.
- Inclined plate metering devices have cups or cells around the periphery that pass through a seed reservoir fed under a baffle from hopper.

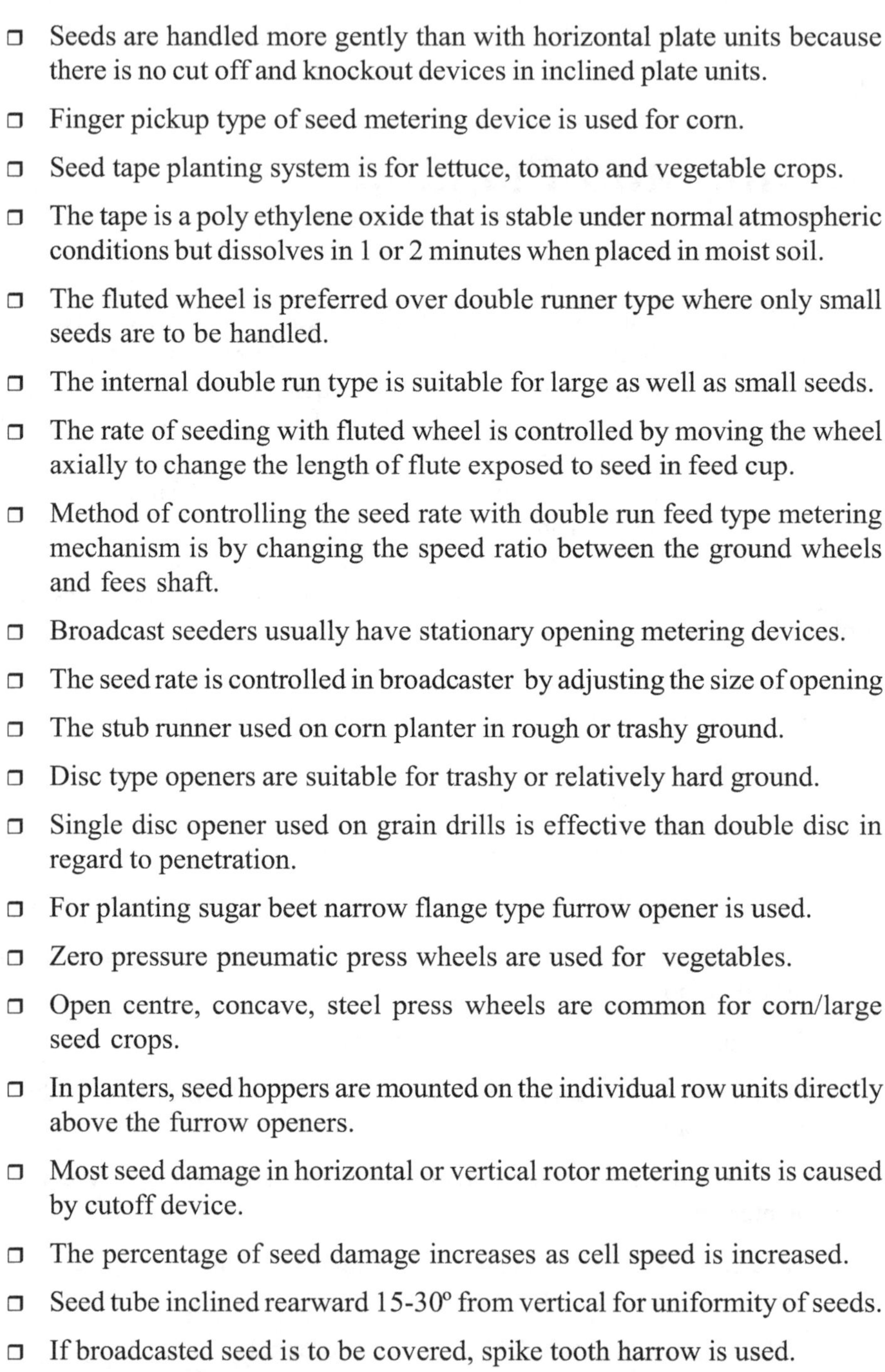

- Seeds are handled more gently than with horizontal plate units because there is no cut off and knockout devices in inclined plate units.
- Finger pickup type of seed metering device is used for corn.
- Seed tape planting system is for lettuce, tomato and vegetable crops.
- The tape is a poly ethylene oxide that is stable under normal atmospheric conditions but dissolves in 1 or 2 minutes when placed in moist soil.
- The fluted wheel is preferred over double runner type where only small seeds are to be handled.
- The internal double run type is suitable for large as well as small seeds.
- The rate of seeding with fluted wheel is controlled by moving the wheel axially to change the length of flute exposed to seed in feed cup.
- Method of controlling the seed rate with double run feed type metering mechanism is by changing the speed ratio between the ground wheels and fees shaft.
- Broadcast seeders usually have stationary opening metering devices.
- The seed rate is controlled in broadcaster by adjusting the size of opening
- The stub runner used on corn planter in rough or trashy ground.
- Disc type openers are suitable for trashy or relatively hard ground.
- Single disc opener used on grain drills is effective than double disc in regard to penetration.
- For planting sugar beet narrow flange type furrow opener is used.
- Zero pressure pneumatic press wheels are used for vegetables.
- Open centre, concave, steel press wheels are common for corn/large seed crops.
- In planters, seed hoppers are mounted on the individual row units directly above the furrow openers.
- Most seed damage in horizontal or vertical rotor metering units is caused by cutoff device.
- The percentage of seed damage increases as cell speed is increased.
- Seed tube inclined rearward 15-30° from vertical for uniformity of seeds.
- If broadcasted seed is to be covered, spike tooth harrow is used.
- Centrifugal type broad caster also known as an endgate seeder.

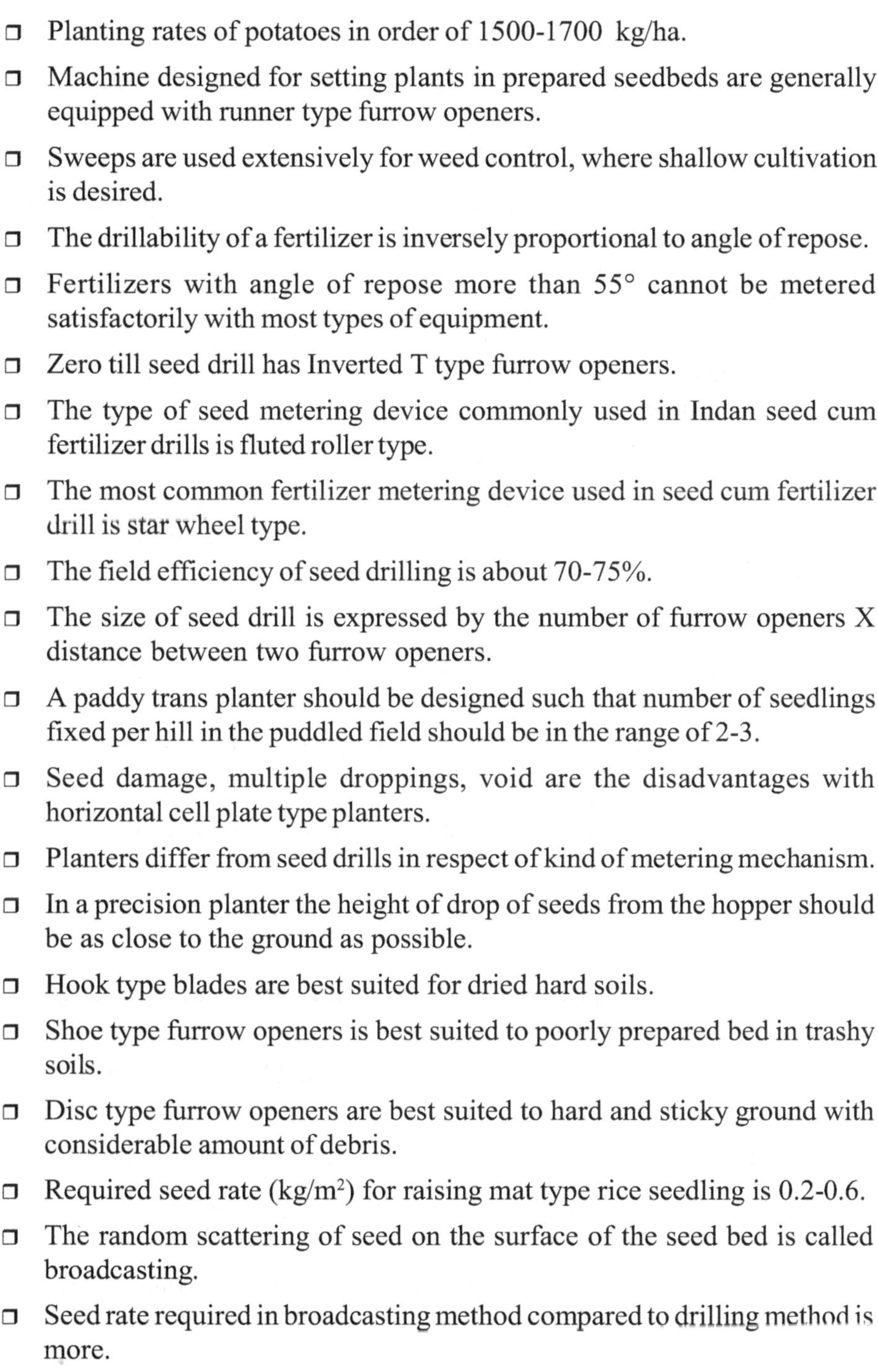

- Planting rates of potatoes in order of 1500-1700 kg/ha.
- Machine designed for setting plants in prepared seedbeds are generally equipped with runner type furrow openers.
- Sweeps are used extensively for weed control, where shallow cultivation is desired.
- The drillability of a fertilizer is inversely proportional to angle of repose.
- Fertilizers with angle of repose more than 55° cannot be metered satisfactorily with most types of equipment.
- Zero till seed drill has Inverted T type furrow openers.
- The type of seed metering device commonly used in Indan seed cum fertilizer drills is fluted roller type.
- The most common fertilizer metering device used in seed cum fertilizer drill is star wheel type.
- The field efficiency of seed drilling is about 70-75%.
- The size of seed drill is expressed by the number of furrow openers X distance between two furrow openers.
- A paddy trans planter should be designed such that number of seedlings fixed per hill in the puddled field should be in the range of 2-3.
- Seed damage, multiple droppings, void are the disadvantages with horizontal cell plate type planters.
- Planters differ from seed drills in respect of kind of metering mechanism.
- In a precision planter the height of drop of seeds from the hopper should be as close to the ground as possible.
- Hook type blades are best suited for dried hard soils.
- Shoe type furrow openers is best suited to poorly prepared bed in trashy soils.
- Disc type furrow openers are best suited to hard and sticky ground with considerable amount of debris.
- Required seed rate (kg/m^2) for raising mat type rice seedling is 0.2-0.6.
- The random scattering of seed on the surface of the seed bed is called broadcasting.
- Seed rate required in broadcasting method compared to drilling method is more.
- The dibbling is mostly used for sowing vegetables.

- Dibbler is a planter.
- The most positive seed metering mechanism used in seed drills is fluted roller type.
- The furrow opener which works well in trashy soil is shoe type.
- The part of seed drill that conveys seed or fertilizer from delivery tube to furrow is called boot.
- The device used to sow seeds which are larger in size and cannot be sown by usual seed drills is planter.
- In a seed metering mechanism, only one seed is allowed at a time in flat drop mechanism.
- The seeding behind the plough is done by a device known as Molobansa.
- The hopper design to plant accurately meter all the varieties of maize and all types of seeds from tomato to beans is duplex hopper, Richmond hopper.
- The metering mechanism suitable for metering small and large seeds is internal double run type.
- In cell feed type metering mechanism, cell fill is affected by size of seeds, speed of cells and size of cells.
- Recommended capacity of seed hopper for manually operated seeders is 2-10 litre.
- Recommended capacity of seed hopper for animal operated seed drills is 10-60 litre
- Recommended capacity of seed hopper for tractor mounted (small size) seed drill is 100-150 litre
- Recommended capacity of seed hopper for tractor mounted (large size) seed drill is 200-300 litre.
- The metering mechanism which is most suitable to handle small seeds is fluted wheel mechanism.
- Seed damage due to cut-off is not a problem in inclined plate planter.
- Seed rate can be changed by changing the speed ratio between ground wheels and metering shafts.
- The drillability of a fertilizer is inversely proportional to angle of repose of fertilizer.

5

Plant Protection: Spraying and Dusting

Sprayer is a machine to apply fluids in the form of droplets. Sprayer is used for the following purpose.

- Application of herbicides to remove weeds.
- Application of fungicides to minimize fungus diseases.
- Application of insecticides to control insect pests.
- Application of micro nutrients on the plants.

Chemical application is an important aspect of the modern agriculture to save crop beset by insects, fungus, virus and other parasites, as well as weeds, which adversely affect a plant growth. Chemical control is still a very effective control method for most insect, pests, weeds and diseases, despite intensive research in to alternative methods. Duster is a machine to apply chemicals in dust form, which makes use of air stream to carry pesticides in finely divided dry form onto the plants. A duster essentially consists of hopper, agitator, feed control, fan or blower and delivery tube and nozzle.

- Sprayers can be classified into three groups i.e. high volume sprayer, low volume sprayer and ultra-low volume sprayer.
- High volume sprayers are those where more than 400 litres of spray liquid per hectare.
- If spray volume is in between 5 to 400 litres per hectare, these sprayers are called as low volume sprayers.
- Sprayers applying less than 5 litres per hectare are called as ultra-low volume sprayers.
- Hollow-Cone Nozzle is commonly used for spraying insecticides and fungicides.
- Solid-Cone Nozzle produces a solid-cone spray pattern and is used for spot spraying of herbicides.

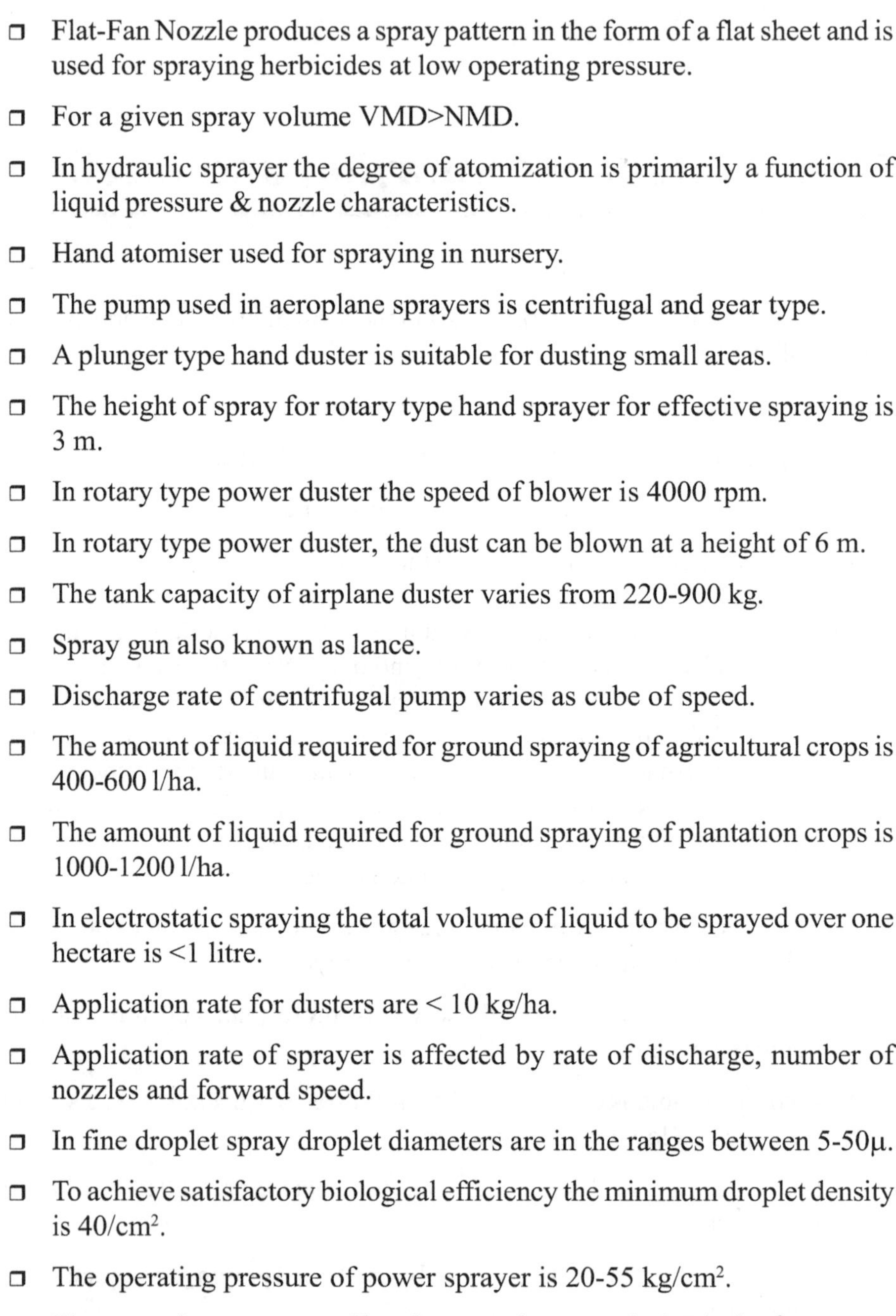

- Flat-Fan Nozzle produces a spray pattern in the form of a flat sheet and is used for spraying herbicides at low operating pressure.
- For a given spray volume VMD>NMD.
- In hydraulic sprayer the degree of atomization is primarily a function of liquid pressure & nozzle characteristics.
- Hand atomiser used for spraying in nursery.
- The pump used in aeroplane sprayers is centrifugal and gear type.
- A plunger type hand duster is suitable for dusting small areas.
- The height of spray for rotary type hand sprayer for effective spraying is 3 m.
- In rotary type power duster the speed of blower is 4000 rpm.
- In rotary type power duster, the dust can be blown at a height of 6 m.
- The tank capacity of airplane duster varies from 220-900 kg.
- Spray gun also known as lance.
- Discharge rate of centrifugal pump varies as cube of speed.
- The amount of liquid required for ground spraying of agricultural crops is 400-600 l/ha.
- The amount of liquid required for ground spraying of plantation crops is 1000-1200 l/ha.
- In electrostatic spraying the total volume of liquid to be sprayed over one hectare is <1 litre.
- Application rate for dusters are < 10 kg/ha.
- Application rate of sprayer is affected by rate of discharge, number of nozzles and forward speed.
- In fine droplet spray droplet diameters are in the ranges between 5-50μ.
- To achieve satisfactory biological efficiency the minimum droplet density is 40/cm^2.
- The operating pressure of power sprayer is 20-55 kg/cm^2.
- The operating pressure of hand operated sprayer is 1-7 kg/cm^2.
- Fan spray and hollow cone nozzles on boom type field sprayer have angle of 60-95 degrees.

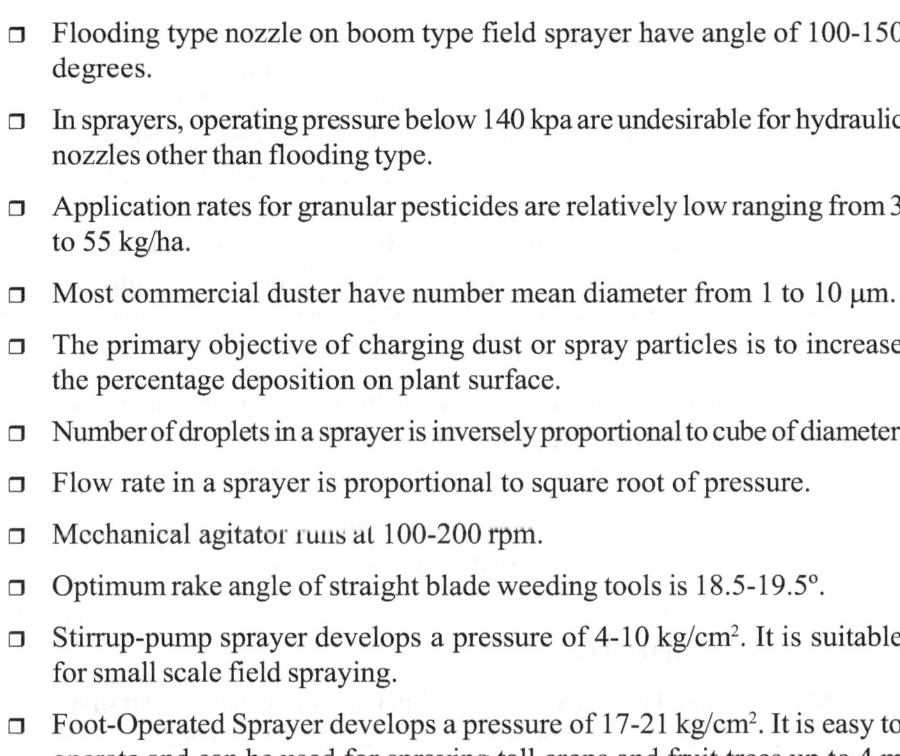

- Flooding type nozzle on boom type field sprayer have angle of 100-150 degrees.
- In sprayers, operating pressure below 140 kpa are undesirable for hydraulic nozzles other than flooding type.
- Application rates for granular pesticides are relatively low ranging from 3 to 55 kg/ha.
- Most commercial duster have number mean diameter from 1 to 10 μm.
- The primary objective of charging dust or spray particles is to increase the percentage deposition on plant surface.
- Number of droplets in a sprayer is inversely proportional to cube of diameter.
- Flow rate in a sprayer is proportional to square root of pressure.
- Mechanical agitator runs at 100-200 rpm.
- Optimum rake angle of straight blade weeding tools is 18.5-19.5°.
- Stirrup-pump sprayer develops a pressure of 4-10 kg/cm^2. It is suitable for small scale field spraying.
- Foot-Operated Sprayer develops a pressure of 17-21 kg/cm^2. It is easy to operate and can be used for spraying tall crops and fruit trees up to 4 m height.
- Rocking Sprayer can develop a pressure of 14-18 kg/cm^2, but in some cases it can develop up to 36 kg/cm^2.
- Blower Sprayers: They are also called mist sprayers.
- Agitator: A device placed inside the tank or hopper, which mechanically stirs the liquid / dust to avoid choking and to smoothly feed to the aperture.
- Centrifugal energy sprayer: A sprayer, which uses the centrifugal force for atomization of liquid droplets.
- Coarse spray: The distribution of droplets having volume median diameter more than 400 micron.
- Concentration: The amount of active ingredient contained in a pesticide formulation. It is expressed as a percent or mass per unit volume (mg active ingredient per liter).
- Contact pesticide: The pesticide, which enters the pests' body through contact when it comes in contact with treated surfaces of the foliage, stem or target area.

- Drift: Drift is of particular concern when deposits of harmful chemicals must be avoided or minimized in the fields nearby the area where chemical is used. The parameters affecting drift are rate of fall of particles, initial height of fall, wind velocity and direction, atmospheric stability and other meteorological factors. Evaporation of water or other volatile materials from droplets reduces the droplet size and thus adversely affects the deposition efficiency and drift. Employing atomizing devices that produce spray having large volume mean diameters can minimize drift. Drift is more serious with dust than with sprays because of smaller particle size.
- Droplet Density (DD): The number of droplets per unit area of leaf surface is called droplet density. Droplet density of 20-50 droplets of pesticide per square centimeter is most effective against cotton pests.
- Hydraulic energy sprayer: A sprayer, which uses a liquid pump to supply the liquid under pressure to a nozzle for atomization.
- Knapsack sprayer: A sprayer, which can be mounted on the back of an operator for spraying.
- Mass Median Diameter (MMD): The mass median diameter of a droplet is numerically the same as volume median diameter (VMD) of the droplet.
- Number Median Diameter (NMD): It is an average diameter of the droplets without any reference to their volume. The diameter corresponding to 50% cumulative number curve will give NMD.
- Spray angle: The angle subtended at the final orifice by the edges of the spray pattern. It is also called as cone angle.
- Spray lance: A hand held tube through which, the liquid passes to a nozzle after being released from cut off device.
- Spread factor: The ratio of the diameter after incidence to the diameter before incidence is called spread factor for a particular object and liquid spread. It is greater than unity because after impingement the droplet spreads.
- ULV sprayer: A gaseous or centrifugal energy sprayer used to apply about 1 to 5 liters of spray liquid per hectare.
- Uniformity Coefficient (UC): The ratio of VMD and NMD is called uniformity coefficient. It indicates the range of size of droplets. The more uniform the size, the nearer the ratio is to unity.

- Volume Median Diameter (VMD): This parameter is usually employed to describe the droplet spectrum. It is droplet diameter that satisfies the condition that half the spray volume is of droplets smaller than a droplet whose diameter is VMD and other half of spray volume contains larger droplets. This can be obtained from a volume cumulative curve of droplet size spectrum. It is the diameter corresponding to 50% cumulative. Droplets in the range of 120-300 micron VMD were found efficacious.

6

Harvesting and Threshing

It is the operation of cutting, picking, plucking and digging or a combination of these operations for removing the crop from under the ground or above the ground or removing the useful part or fruits from plants.

Thresher is a machine to separate grains from the harvested crop and provide clean grain without much loss and damage. During threshing, grain loss in terms of broken grain, un-threshed grain, blown grain, spilled grain etc. should be minimum. Bureau of Indian Standards has specified that the total grain loss should not be more than 5 per cent, in which broken grain should be less than 2 per cent. Clean un-bruised grain fetch good price in the market as well as it has long storage life.

It is a machine designed for harvesting, threshing, separating, cleaning and collecting grains while moving through standing crops. Bagging arrangement may be provided with a pick up attachment. The main functions of a combine are: (i) Cutting the standing crops (ii) Feeding the cut crops to threshing unit (iii) Threshing the crops (iv) Cleaning the grains from straw (v) collecting the grains in a container.

Harvester	Action
Sickle	Manual Harvester
Mower	Cutting of forage
Reaper	Cutting of grain crop
Reaper Binder	Cutting and tying of grain crop
Windrower	Cutting and delivering the harvest in an uniform manner

- For proper operation, the reel speed index of combine harvester is kept 1.25-1.50.
- Cutter bar speed of reaper for harvesting cereal crops is about 1-1.4 m/s.
- The ratio of peripheral velocity to the speed of travel of combine is called reel index.

- The purpose of registration in mower is to run the mower at minimum power.
- The useful life of combine(mounted and drawn type) is 7 years.
- The Olpad thresher is mostly used for threshing wheat.
- Olpad thresher is bullock drawn type and operated by a pair of bullocks.
- The stripping method of threshing is used for paddy.
- Loop type thresher is generally used in paddy thresher.
- A thresher which is not equipped with aspirator is called drummy thresher.
- Winnower is the machine consisting of one or two sieves.
- A thresher causes more damage if the speed is increased.
- Type of cylinders used on power wheat threshers is spike tooth/ hammer mill.
- The output capacity of hand beating method is about 17-20 kg/h.
- The drum diameter of rotary type thresher is 43 cm.
- For safety the feeding chute of a power thresher should be covered upto 45cm.
- For satisfactory threshing, the clearance between cylinder and concave should be more.
- Uneven and excessive feeding of crops into cylinder, causes the problem of blowing of grain with straw.
- As per BIS recommendations, the rpm of the threshing drum for wheat crop is 550-1100 rpm.
- The recommended speed of threshing drum for paddy is 675 to 1000 rpm.
- The size of power thresher is determined by the width of cylinder and straw rack.
- In power thresher, a semi-circular disc is inserted between cylinder and straw thrower to achieve higher threshing efficiency.
- The average threshing capacity of bullock drawn Olpad thresher per day is 562 kg.
- A man can thresh average grain in day is about 19 kg.
- A cutter bar used in combine is of mower type.
- Optimum operating cylinder speed for wheat threshing using power operated thresher is 1100-1200 rpm.

- Optimum speed for grains like gram, pea etc., using power thresher is 300-400 rpm.
- Recommended cylinder speed for threshing wheat crop is 20-25 m/s.
- The cutting platform of combine operates at a height of 7.5-9 cm above the ground.
- The suitable clearance between reel and cutter bar for all purposes is 12.5-25 cm.
- The size of combine is determined by the width of harvest.
- Threshing cylinder speed is given by ðdn.
- In axial flow paddy thresher a spike tooth cylinder is preferred over a raspbar cylinder because it has more positive feeding action, requires less power, does not plug easily.
- Cutting speed of cuter bar of a mower is 2 m/s.
- The cutting width of a small size self-propelled combine ranges from 1.5-2.1 m.
- For wheat the recommended speed of threshing drum is 20-30 m/s.
- For paddy the recommended speed of threshing drum is 16-25 m/s.
- The speed of hand operated winnowing fan varies from 200-350 rpm.
- A conventional pitman drive mower has crank speed of 850-1100 rpm.
- The size of cutter bar of self-propelled combine varies from 2-4 m.
- The operating speed of combine in field varies from 2-6.4 kmph.
- The reel most commonly used in combine harvesters is pickup type reel.
- The rack loss should remain 0.2-0.4 %.
- The mowers are designed to cut grasses.
- The width of cut of animal drawn reaper is 900 mm.
- Tractor front mounted reaper is operated at a speed of 3 kmph.
- The conventional animal drawn mover has a stroke length of 7.5 cm.
- The hand dropper was developed in japan.
- The cutting height of reaper binder from ground is about 10 cm.
- In the reaper shoe is provided at each end of cutter bar.
- Cutter bar of reaper is made of high grade steel.
- Power requirement of spike tooth thresher at no load should not more than 20 % of rated power.

- A reciprocating type mower is fixed with reciprocating type cutter bar.
- Axial flow thresher uses rasp bar type cylinder.
- The output capacity (kg/kW-h) of wheat thresher should not be less than 85 %.
- The threshing efficiency of wheat thresher should not be less than 96%.
- The cleaning efficiency of wheat thresher should not be less than 95%.
- Cutting speed of impact type cutter is 46-56 m/s.
- Flail mower uses high speed swinging knives.
- In combine harvester the peripheral speed of the cylinder ranges between 18-33 m/s.
- During the operation of the cutter bar mower, the knife edge traces a straight path in relation to field surface.
- In a combine harvester, the peripheral speed of cylinder ranges between 1500-1800 m/min.
- Hay may contain 70-80 % of water when first harvested.
- For storage hay must be dried either naturally or artificially to a safe moisture content of about 20-25 %.
- A rotary cutter has knives rotating in a horizontal plane where as a flail shredder has knives rotating in vertical planes parallel with direction of travel.
- Rotary cutters with effective width of 1 to 2.1 m have single rotor usually 2 knives on ends of radial arm .
- Peripheral speeds of rotary cutters range from 51-76 m/s.
- Flail shredders have free swinging knives or flails with a peripheral speed of 46-56 m/s.
- Pull type flail mower are generally offset to tractor wheels run on cut hay rather than on standing crop.
- A mower cutter bar must be able to cut satisfactorily from 25-30 mm above ground level upto 100 mm.
- The guard spacing is almost universally 76.2 mm.
- Height of cut is changed by adjustable shoes at the end of cutter bar.
- The knife may be driven through a long pitman attached to the knife by means of a ball & socket joint.

- Crank speeds on most mowers are 850-100 rpm for conventional pitman drives and it is 1000-1200 rpm for dynamically balanced mowers.
- Mounting the drive unit directly on the inner shoe of the cutter bar eliminates the problem of maintaining alignment and registration.
- Inertia force varies as square of the crank speed.
- The optimum inclination of star wheel inclination of reaper windrower is 32°.
- The optimum velocity ratio of reaper cutter bar to machine forward speed is 0.75-1.
- The tractor drawn semi-mounted or mounted type mowers are operated by the P.T.O. shaft. In this case, the cutting mechanism is driven independently of the forward speed of the mower. A shaft is connected with the PTO shaft, which drives a V pulley with the help of a universal joint. The V pulley rotates another smaller pulley on the crank shaft of the machine and reciprocating motion is transmitted to the cutter bar.
- The length of the cutter bar stroke is 7.5 cm.
- Knife clips are placed together with wearing plates to absorb the rearward thrust of the crop to the knife.
- Knife clips hold the knife sections down against ledger plates but allow it to move freely.
- Ledger plate is a hardened metal, inserted in a finger over which knife sections move to give a scissor like cutting action.
- The crank wheel to which the pitman is attached also act as a small flywheel to help carry the knife through its cutting stroke.
- Shoe is provided on each end of the cutter bar to regulate the height of cut above the ground.
- The inner shoe is larger in section and is placed at the inner end of the cutter bar.
- The outer shoe is placed at the outer end and is smaller in section.
- The inner shoe has a larger area of contact with the ground than the outer one.
- Grass board is provided at the outer end of the mower, which causes the cut plants to fall towards the cut material.
- Generally, 2 cm lead per metre length of cutter bar is adequate.

- The operation of cutting a plant is achieved by four different actions.
  - Slicing action with a sharp smooth edge,
  - Tearing action with a rough, serrated edge,
  - High velocity single element impact with sharp or dull edge, and A two element scissor type action.
- Manual harvesting involves slicing and tearing actions that result in plant structure failure due to compression, tension or shear.
- Single element impact cutting is an economical method of cutting unrestrained vegetation and has been widely used in rotary lawn mowers, forage choppers, and some tractor mounted cutter bar.
- Usually a single element, sharp edged blade requires a velocity of about 10 m/s for impact cutting.
- A dull edged, single element blade requires a velocity of about 45 m/s.
- The two element scissors action is the most widely used for harvesting agricultural crops.
- The reciprocating cutter bars that are commonly used for harvesting paddy/ wheat.
- The inclined angle between the cutting edges is about 38 degrees.
- The serrated blades permit a larger inclined angle because the plants cannot easily slip between the two cutting edges.
- Reciprocating cutter bars do an excellent job of harvesting but are characterized by the high energy, losses, short dynamic imbalance, and restricted operating speeds.
- The forged end of the blade for fixing the handle is called tang.
- The plain or serrated edge in the inner side of the blade is called cutting edge.
- Protective metallic bush fitted at the junction of the blade and the handle to keep the tang tight in the handle is called ferrule.
- Mower is a machine to cut herbage crops and leave them in swath.
- Bullock drawn mowers: The main axle receives power from one of the transporting wheels. A spur gear mounted on the main axle drives the spur pinion on one end of the countershaft in the gear box. The crank wheel and the pitman are fixed on the outer end of the crankshaft. The

reciprocating (back and forth) motion is transmitted to the pitman, which in turns operates the knife in the cutter bar. The knife is connected to the pitman with a ball and socket joint.

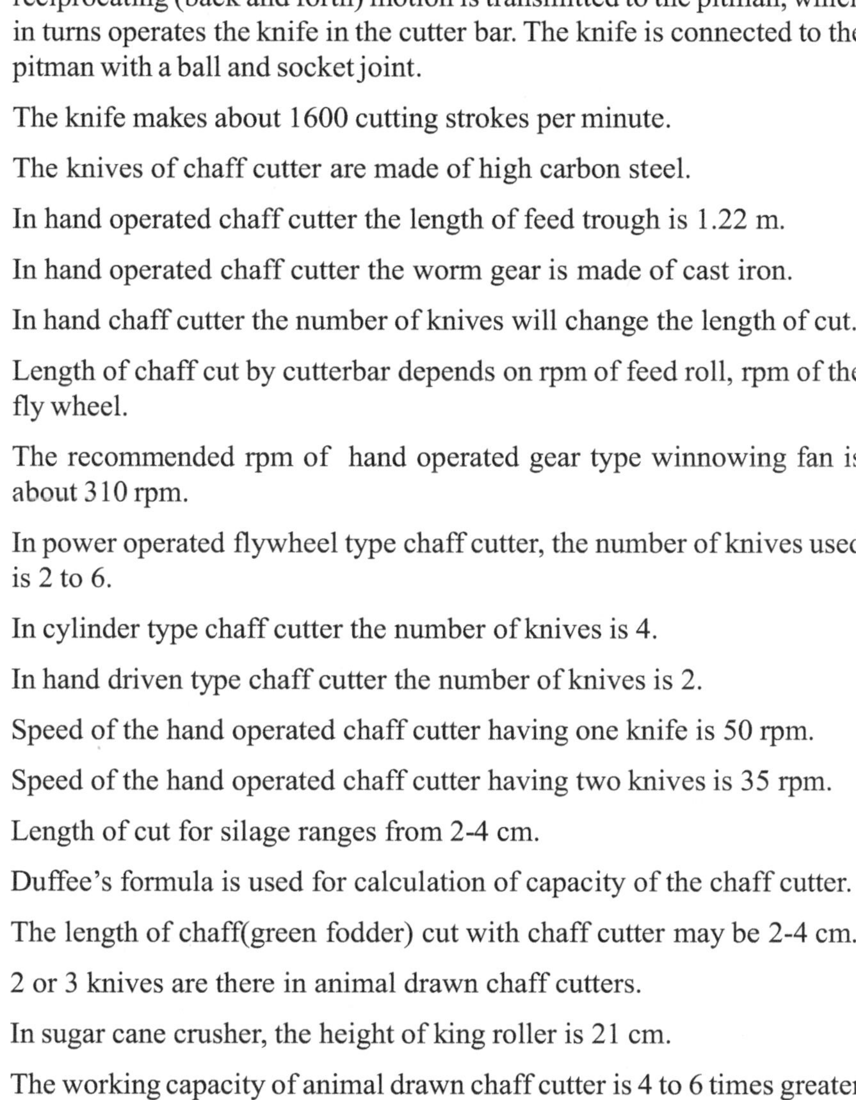

- The knife makes about 1600 cutting strokes per minute.
- The knives of chaff cutter are made of high carbon steel.
- In hand operated chaff cutter the length of feed trough is 1.22 m.
- In hand operated chaff cutter the worm gear is made of cast iron.
- In hand chaff cutter the number of knives will change the length of cut.
- Length of chaff cut by cutterbar depends on rpm of feed roll, rpm of the fly wheel.
- The recommended rpm of hand operated gear type winnowing fan is about 310 rpm.
- In power operated flywheel type chaff cutter, the number of knives used is 2 to 6.
- In cylinder type chaff cutter the number of knives is 4.
- In hand driven type chaff cutter the number of knives is 2.
- Speed of the hand operated chaff cutter having one knife is 50 rpm.
- Speed of the hand operated chaff cutter having two knives is 35 rpm.
- Length of cut for silage ranges from 2-4 cm.
- Duffee's formula is used for calculation of capacity of the chaff cutter.
- The length of chaff(green fodder) cut with chaff cutter may be 2-4 cm.
- 2 or 3 knives are there in animal drawn chaff cutters.
- In sugar cane crusher, the height of king roller is 21 cm.
- The working capacity of animal drawn chaff cutter is 4 to 6 times greater than hand operated chaff cutter.
- The juice extraction capacity of animal drawn sugar cane crusher is 50-70 %.
- The number of rollers in animal drawn cane crusher is 3.

7

Theory of Machine and Machine Design

- A machine is a device which receives some sort of energy and does some work.
- A rigid link is one, which does not undergo any deformation while transmitting motion.
- A flexible link is one which is partly deformed in a manner not to affect the transmission of motion.
- A structure is an assemblage of a number of resistant bodies having no relative motion between them.
- A turning pair has a completely constrained motion.
- Cam and follower includes higher pair.
- The lower pairs are self-closed pair.
- The cam and follower is an example of force closed pair, as it is kept in contact by the forces exerted by spring and gravity.
- When a circle rolls without slipping on the outside of a fixed circle, the curve traced by the point on the circumference of a circle is known as Epicycloid.
- The relative motion caused by the changes in length of the belt when it is passed from the slack side to tight side and vice versa is known as creep of the belt.
- In order to maintain good results with flat belts, the maximum distance between the shafts should not exceed 10 m and the minimum should not be less than 3.5 times the diameter of the larger pulley.
- The friction experienced by a body, due to the motion of rotation as in case of foot step bearings is called pivot friction.
- The friction material used in plate clutch is ferredo.
- The subject theory of machines may be defined as the branch of engineering science, which deals with the study of relative motion between the various parts of a machine, and forces which act on them.

- The two elements of a machine when in contact with each other, are said to form a pair/kinematic pair.
- The method of obtaining different mechanisms by fixing different links in a kinematic chain is known as inversion of mechanisms.
- Kinetics is the branch of theory of machines which deals with the inertia forces which arise from the combined effect of the mass and motion of the machine parts.
- Frame and tine of a cultivator constitutes a link.
- The mechanism of coupling rod of locomotive is also known as Double crank mechanism.
- The mechanism used in shaping machines is Crank and slotted lever Quick Return Motion Mechanism.
- The gear train observed in tractor gear box is Reverted gear train.
- Standard pressure angle of a gear is 14½ to 22 degrees.
- In quarter pulley belt drive the width of the face of the pulley should be greater or equal to 1.4b If b is the width of the belt.
- Circular belts or ropes can be used where Great amount of power can be transmitted.
- In order to have a complete balance of the several revolving masses in different planes. Both the resultant force and couple must be zero.
- The velocity ratio of two pulleys connected by an open belt and crossed belt is Inversely proportional to their diameters.
- For high speed engines, the cam follower should move with Cycloid motion.
- The product of the diametric pitch and circular pitch is equal to ð.
- A kinematic chain is known as a mechanism when One of the links is fixed.
- The two parallel and coplanar shafts are connected by gears having teeth parallel to the axis of the shaft. This arrangement is known as Spur gearing.
- The size of the cam depends upon Base circle.
- In low and moderate speed engine, the cam follower should move with Simple harmonic motion.
- In a gear drive the velocity is increased by 2 times and the Torque is decreased by 2 times the output power Remains same.

- A straight tooth gear having involute profile with 20 degrees pressure angle transmits 20 Nm torque. Pitch diameter of the gear is 24 cm. If only one pair of tooth is engaged at any instant of time, the force experienced by the tooth will be 177.36 N.
- In an epicyclic gear train, an arm carries two wheels A and B having 24 teeth and 30 teeth respectively. If the arm rotates at 100 rpm in the clockwise direction about the centre of the wheel 'A' which is fixed, then the speed of wheel on its own axis is 180 rpm, clockwise.
- The radial distance of a tooth from the pitch circle to the top of the tooth is called Addendum.
- Type of belt drive used when shafts are to be connected at perpendicular to each other is Quarter turn belt drive.
- The balancing of rotating and reciprocating parts of an engine is necessary when it runs at high speed.
- When a point moves along a straight line, it has radial component of acceleration.
- The curve traced by point on the circumference of a circle which rolls without slipping on a fixed straight line is called cycloid.
- Balata belts are not suitable at temperatures above 40°C.
- The gears having velocity of >15 m/s are termed as high velocity gears.
- The circle passing through the top of the meshing gear is known as clearance circle.
- Backlash is the difference between the tooth space and the tooth thickness as measured along the pitch circle.
- When motion between pair is limited to definite direction it is called Completely constrained motion.
- When the two elements of a pair have a line contact, the pair formed is called as higher pair.
- Skew bevel gear means shafts are Non-parallel / Non-intersecting and Non-coplanar.
- The branch of theory of machines which deals with the forces and their effects when the machine parts are at rest is called Statics.
- The law used in four bar chain mechanism is Grashof's law.

- The link which connects crank and lever is called Coupler.
- Porter governor cannot be isochronous.
- In external gearing the large wheel is called Spur wheel.
- When flat faced follower is circular, it is called Mushroom follower.
- Gear mechanism in natural world can be found in insect called Issus.
- A Stepped or cone pulley drive is used to change the speed or driven shaft without changing the speed of driver.
- When a point at the end of link moves with constant angular velocity will have radial component of acceleration.
- The fixed outer element of a turning pair is called Bearing.
- Function of a governor is to regulate engine speed.
- Reciprocal of module is called Diametral pitch.
- Reciprocal of speed ratio is called Train value.
- The two parallel and coplanar shafts are connected by gears having teeth inclined to the axis of the shaft. This arrangement is known as Helical gearing.
- The radial distance of a tooth from the pitch circle to the bottom of the tooth is called Dedendum.
- According to law of gearing, the common normal at the point of contact between a pair of teeth must always pass through Fixed point & Pitch point.
- The belt speed of heavy drive is >22 m/s.
- Slip is generally expressed in %.
- The branch of theory of machines which deals with the forces and their effects while acting upon machine parts is called Dynamics.
- A cam is a rotating machine element which gives reciprocating or oscillating motion to another element known as follower.
- When the contacting end of the follower has a sharp knife edge, it is called a knife edge follower.
- When the contacting end of the follower is a roller, it is called a roller follower.

- When the contacting end of the follower is a perfectly flat face, it is called a flat-faced follower.
- When the contacting end of the follower is of spherical shape, it is called a spherical faced follower.
- When the follower reciprocates in guides as the cam rotates uniformly, it is known as reciprocating or translating follower.
- When the uniform rotary motion of the cam is converted into predetermined oscillatory motion of the follower, it is called oscillating or rotating follower.
- When the motion of the follower is along an axis passing through the centre of the cam, it is known as radial follower.
- When the motion of the follower is along an axis away from the axis of the cam centre, it is called off-set follower.
- Pitch point is a point on the pitch curve having the maximum pressure angle.
- Prime circle is the smallest circle that can be drawn from the centre of the cam and tangent to the pitch curve.
- Lift or stroke is the maximum travel of the follower from its lowest position to the topmost position.
- Light drives are used to transmit small powers at belt speeds upto about 10 m/s, as in agricultural machines and small machine tools.
- Medium drives are used to transmit medium power at belt speeds over 10 m/s but up to 22 m/s, as in machine tools.
- Compound belt drive. A compound belt drive is used when power is transmitted from one shaft to another through a number of pulleys.
- Belt drive with idler pulleys is used with shafts arranged parallel and when an open belt drive cannot be used due to small angle of contact on the smaller pulley.
- The quarter turn belt drive also known as right angle belt drive is used with shafts arranged at right angles and rotating in one definite direction.
- Velocity ratio is the ratio between the velocities of the driver and the follower.
- The tension caused by centrifugal force is called centrifugal tension.
- At lower belt speeds (less than 10 m/s), the centrifugal tension is very small.

- The size of the rope is usually designated by its circumference.
- When the motion between a pair can take place in more than one direction, then the motion is called an incompletely constrained motion.
- A chain having more than four links is known as compound kinematic chain.
- When two links are joined at the same connection, the joint is known as binary joint.
- When three links are joined at the same connection, the joint is known as ternary joint.
- When four links are joined at the same connection, the joint is called a quaternary joint.
- A mechanism with four links is known as simple mechanism, and the mechanism with more than four links is known as compound mechanism.
- Pivot friction. It is the friction, experienced by a body, due to the motion of rotation.
- Sliding friction. It is the friction, experienced by a body, when it slides over another body.
- Rolling friction. It is the friction, experienced between the surfaces which have balls or rollers interposed between them.
- Boundary friction is the friction experienced between the rubbing surfaces, when the surfaces have a very thin layer of lubricant.
- Fluid friction is the friction experienced between the rubbing surfaces, when the surfaces have a thick layer of the lubricant.
- Static friction has any value between zero and limiting friction.
- Coefficient of friction is the ratio of the limiting friction to the normal reaction between the two bodies.
- If the lead of a screw is equal to its pitch, it is known as single threaded screw.
- When equal bevel gears (having equal teeth) connect two shafts whose axes are mutually perpendicular, then the bevel gears are known as mitre gear.
- The phosphor bronze is widely used for worm gears in order to reduce wear of the worms.

- If a circle rolls without slipping on the inside of a fixed circle, then the curve traced by a point on the circumference of a circle is called hypo-cycloid.
- The variations of energy above and below the mean resisting torque line are called fluctuations of Energy.
- The difference between the maximum and the minimum energies is known as maximum fluctuation of energy.
- Coefficient of fluctuation of energy is the ratio of the maximum fluctuation of energy to the work done per cycle.
- The difference between the maximum and minimum speeds during a cycle is called the maximum fluctuation of speed.
- The ratio of the maximum fluctuation of speed to the mean speed is called the coefficient of fluctuation of speed.
- The reciprocal of the coefficient of fluctuation of speed is known as coefficient of steadiness.
- The function of a governor is to regulate the mean speed of an engine.
- Height of the governor is the vertical distance from the center of the ball to a point where the axes of the arms intersect on the spindle axis.
- Equilibrium speed is the speed at which the governor balls, arms etc., are in equilibrium and the sleeve do not tend to move upwards or downwards
- Mean equilibrium speed is the speed at mean position of the balls or the sleeve.
- The effort of a governor is the mean force exerted at the sleeve for a given percentage change of speed.
- The power of a governor is the work done at the sleeve for a given percentage change of speed.
- Epicyclic gear trains are used to transmit high velocity ratios with gears of moderate size in a comparatively less space.
- Plain roller bearings have a high radial load capacity but no end thrust.
- Plain roller bearings are most suitable for low speed applications such as on axle of a grain drill.
- Tapered roller bearings are designed to take both radial & thrust loads, the relative capacity depends on the amount of taper.
- Life of an anti-friction bearing in terms of stress cycle vary as inverse cube of load.

- Life of an anti-friction bearing in hours inversely proportional to the rotational load.
- V belts are operated at 33 m/s.
- The agricultural machinery applications seldom exceeds 15 m/s.
- V belts are not suitable for heavy loads at low speed.
- A single belt on an implement offers drive to several components in an arrangement known as serpentine drive.
- Life expectancy of 1000-2000 h is adequate for most farm machinery applications.
- Service factor for agricultural machinery is 1.2-1.5.
- If the ratio between tight side & slack side tensions is to great, belt slippage will be excessive.
- Load capacity of a chain is based on the rate of wear rather than its ultimate strength.
- Lesser the teeth on sprocket greater the speed variation.
- Detachable link chain length is 2-2.5 m/s.
- Standard pitch roller chain length is 0.5-2 m/s.
- Shear devices are most used where overloads are rather infrequent.
- Jump clutch is more suitable than shear devices where overloads may occur rather frequently.
- The type of universal joint generally used on agricultural machinery is known as cardon or hook joint.
- The peak lag and peak lead occurs when average angle of rotation of the two shafts is 45° or 135°.
- Bearing materials are made of babbit material.
- When V belt drive is used as safety device, its performance is affected by variation.
- Lesser is the no of teeth on sprocket, the variation in the speed of chain drive is greater.
- Limiting value of speed ratio of standard pitch roller chain is 10:1.
- No. of teeth on sprocket should not be less than for high speed drive to give minimum speed variation is 17-18.

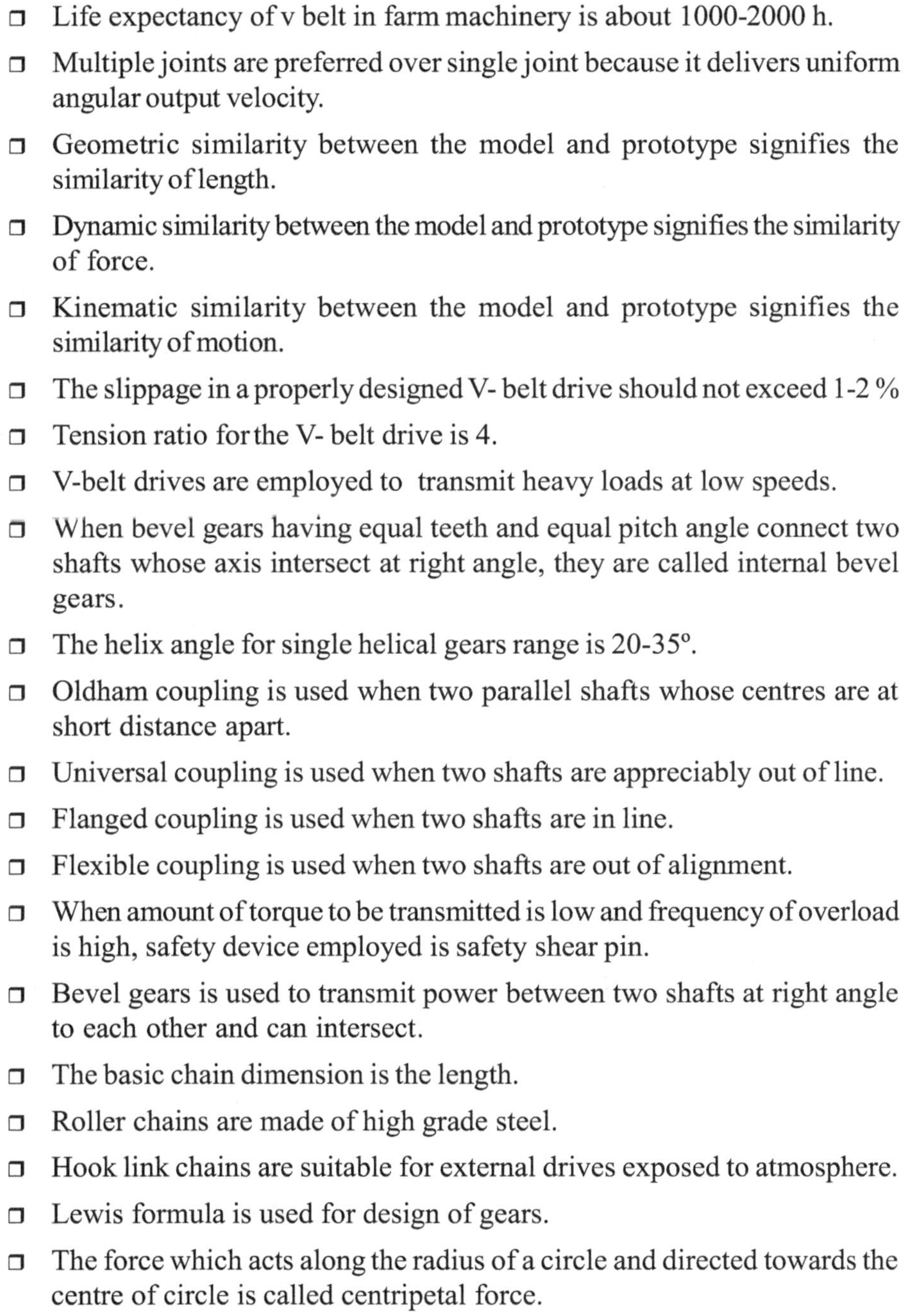

- Life expectancy of v belt in farm machinery is about 1000-2000 h.
- Multiple joints are preferred over single joint because it delivers uniform angular output velocity.
- Geometric similarity between the model and prototype signifies the similarity of length.
- Dynamic similarity between the model and prototype signifies the similarity of force.
- Kinematic similarity between the model and prototype signifies the similarity of motion.
- The slippage in a properly designed V- belt drive should not exceed 1-2 %
- Tension ratio for the V- belt drive is 4.
- V-belt drives are employed to transmit heavy loads at low speeds.
- When bevel gears having equal teeth and equal pitch angle connect two shafts whose axis intersect at right angle, they are called internal bevel gears.
- The helix angle for single helical gears range is 20-35°.
- Oldham coupling is used when two parallel shafts whose centres are at short distance apart.
- Universal coupling is used when two shafts are appreciably out of line.
- Flanged coupling is used when two shafts are in line.
- Flexible coupling is used when two shafts are out of alignment.
- When amount of torque to be transmitted is low and frequency of overload is high, safety device employed is safety shear pin.
- Bevel gears is used to transmit power between two shafts at right angle to each other and can intersect.
- The basic chain dimension is the length.
- Roller chains are made of high grade steel.
- Hook link chains are suitable for external drives exposed to atmosphere.
- Lewis formula is used for design of gears.
- The force which acts along the radius of a circle and directed towards the centre of circle is called centripetal force.

8

Tractor and Its Control System

Tractor is a self-propelled power unit having wheels or tracks for operating agricultural implements and machines including trailers. Tractor engine is used as a prime mover for active tools and stationary farm machinery through power take-off shaft (PTO) or belt pulley.

Power tiller is a prime mover in which the direction of travel and its control for field operation is performed by the operator walking behind it. It is also known as hand tractor or walking type tractor. The concept of power tiller came in the world in the year 1920. Japan is the first country to use power tiller on large scale. Power tiller was first introduced in India in the year 1963. Power tiller is a walking type tractor. The operator walks behind the power tiller, holding the two handles of power tiller in his own hands. Power tiller may be called a single axle walking type tractor, though a riding seat is provided in certain designs. In agricultural applications it is often required to transmit the mechanical power from its source to the point of application. The main sources of power for agricultural machines, i.e., the diesel engine for self-propelled machines and the electrical motor for many stationary machines are primarily rotary power and are transmitted by means of belts and chains. Tractor power is utilized by traction, power-take-off drives (PTO) by fluid power.

- The drawbar power output is always less than the PTO power output because of drive wheel slippage, tractor rolling resistance.
- T&T coefficient for four wheel tractor will be higher than those for two wheel drive tractor.
- T&T coefficient for track type tractor would be at least 0.8-0.85 on firm soil, 0.7-0.75 for soft soils.
- Axle power is considered to be 96% of PTO power.
- The hitch point is to be directly beneath the extended centre line of the PTO shaft.
- Hydraulic pumps/motors have 75-90 % efficiency.

- Most implements with hydraulic system and most medium or small size tractors have constant flow system.
- Rated pressure for constant flow variable pressure systems on tractor are mostly 8-14 MPa.
- Rated pressure for constant flow constant pressure systems on tractor are mostly 12-18 MPa.
- System relief valves are normally set above 25 % higher than the rated pressure.
- Variable flow type variable pressure systems are basically closed loop arrangements used in hydrostatic propulsion drive.
- Axial piston motors are used in hydrostatic propulsion drives for self-propelled machinery.
- Spool type valves are used as directional control valves.
- Spring loaded ball or poppet type check valves are used extensively in hydraulic systems.
- Most cylinder controls on self-propelled implements are of nudging type.
- Draft sensing meter is always in compression.
- Upper link of tractor is subjected to compression.
- Lower link of the tractor is subjected to tension.
- The centre of pull of the tractor is generally considered to be midway between the rear wheels.
- Dimensions are included for 4 hitch categories for different ranges of maximum drawbar power.
  - Category-I 15-35 Kw (20-45 hp).
  - Category-II 30-75 Kw (40-100 hp).
  - Category-III 60-168 Kw (80-225hp).
  - Category-IV 135-300 Kw (180-400 hp).
- In an epicyclic gear speed reduction unit, the no ratio of the no of teeth of the annular gear to the sun gear is 4:1.
- The virtual hitch point for a directional implement is usually located between two lower links.

- One of the main advantage of disc brakes over drum brakes is that disc braked require less force to stop the vehicle.
- Higher tractive efficiency occurs on a dry and hard field.
- The draft for most pull type non tillage implements is in the form of rolling resistance.
- Quick attaching couplers for three point hitch have been developed to facilitate the attachment of implements that are too heavy to be nudged into position by one man.
- Automatic draft control is a type of restrained link system in which the depth of the implement is automatically adjusted to maintain a preselected constant depth.
- Upper link draft sensing was employed on all early automatic draft control systems upper link is always in compression when tool is on ground.
- The stability of 4 wheel drive tractor up a hill can be improved by decreasing the the height of line of action of drawbar pull and increasing the wheel base.
- The volumetric efficiency of a turbo charged diesel engine is 90-100%.
- Tractors only one third of the heating value of fuel is converted into useful work.
- Maximum efficiency will be obtained with high compression ratio/ early fuel cutoff for same compression ratio or same heat input diesel cycle is less efficient than otto cycle.
- Butane boils at 0┐ c at atmospheric pressure.
- Normal heptane is the anti-knock fuel used as a reference fuel in knock testing.
- Alpha methyl naphthalene is used as a reference fuel in testing diesel fuel
- API gravity of water is 10.
- Hydrometer is used for testing gravity of fuels.
- Thermal efficiency of a tractor engine is about 33 %.
- The volumetric efficiency of a naturally aspirated diesel engine is 85-90%.
- Otto, 1876 invented the first successful internal combustion engine.

- J. froelich is the inventor of the tractor.
- The average size of farm holding in India is 1.15 ha.
- The first tractor company introduced air cooled engine in india was EICHER.
- Sound level of tractor for 8 hours operator exposure should not exceed 90 dB.
- When ratio of tractor chasis frequency to set natural frequency is more than 2 the viscous damping increases the vibration transmissibility.
- Operator tolerance to vertical vibration for 8 hours duration is 0.315 m/s^2 (RMS).
- Size of tyre is normally expressed as tyre section width X outer diameter.
- The size of tractor tyre may be represented as section thickness X rim diameter.
- Ply rating of tyre represents the load bearing capacity.
- Oil to be used in engine for lubrication purpose is SAE-30.
- Lubricant used in gear box is SAE-90.
- The grade oil used in winter seasons is SAE-30.
- The grade oil used in summer is SAE-40.
- The voltage in spark plug at the time of spark is 20000 V.
- The standard test track for drawbar test should have good adhesion characteristics with a coefficient of traction in the range of 0.6-0.7.
- Compression pressure of DI tractor engine ranges from 262-362 Kpa.
- Type of fuel injection nozzle generally used in Indian tractors is multiple hole type.
- In I head engine both valves are provided in the cylinder head.
- Engine main bearing tightening torque is usually kept 24-45 kg-m.
- Injection of fuel in CI engine commences at an angle of 5-10° before TDC.
- Gear generally used in transmission box are helical & spur gears.
- Connecting rod of four stroke engine is subjected to compressive load and tensile load.

- Connecting rod of two stroke engine is subjected to compressive load only.
- Radial ply tyres can be operated at more inflation pressure than the biased ply tyres.
- Typical values for coefficient of rolling resistance on settled tilled soil are 0.15-0.35.
- Air standard engine cycle efficiency of petrol engine in comparison to diesel engine got the same compression ratio is higher.
- Most preferred firing order of six cylinder engine is 1-5-3-6-2-4.
- For smoother revolution of crank shaft the best firing order for four stroke, 4 cylinder engine is 1-3-4-2.
- In three cylinder inline engine, the power impulse occurs after every 240° of crank shaft revolution.
- Suggested difference in mass to power ratio between ballasted and unballasted tractor is of the order 6 kg/pto Kw.
- Weight per unit of drawbar power is generally ranges between 50-120 kg/Kw.
- Dual wheels have disadvantage of concentrating load over an area.
- Tandem wheels are preferred over dual wheels because they tends to reduce load on an area by 50 %.
- Weight transfer from front to rear wheels is advantageous in case of 2 wheel drive tractor.
- The lower value of the ratio of drawbar height to length of tractor is better as variation in weight transfer is decreased.
- When the valves are carried in the engine block at one side of the cylinder is called L head.
- Turbo chargers or super chargers increase the power of the engine by 10-15 %.
- Turbo chargers are driven by exhaust gases of the engine.
- Super chargers are driven by direct connection from engine power.
- Chemically correct air fuel ratio of gasoline engine is 15.05:1.
- Recommended engine oil for use in winter is SAE-30.

- Cooling system handles about one third of the heat provided by the fuel.
- Specific gravity of battery electrolyte at fully charged condition is 1.280.
- In direct injection of diesel engines ignition delay is shorter.
- A cylinder block is normally made of cast iron.
- In a two port two stroke engine with crank case compression scavenging will be poor, efficiency will be low, lubrication will be difficult.
- The compression ratio of the diesel engine are in the ratio 14:1 to 20:1.
- Maximum noise level from a tractor near the operator ear should not exceed 85dB.
- A connecting rod of a tractor engine is made of forged steel.
- Nozzles of diesel engines operate usually at a pressure of 125-175 kg/cm^2.
- Mounted implements are attached to tractor by means of three hitch points.
- The mounted implements are directly controlled by the tractor steering and are fully supported by the tractor.
- Trailed implements are connected to tractor by means of only one hitch point.
- The trailed implements are pulled and guided by the tractor but not fully supported by the tractor.
- Most of the indigenous tractors are fitted with a sliding mesh or sliding-cum-constant mesh-type gear box.
- The power available at the crank shaft is called power output or brake power.
- The brake power of an engine is measured by an apparatus called dynamometer.
- As the compression ratio is high, the diesel engines are heavier and costlier.
- The running cost of diesel engine is low because of the lower cost of diesel.
- A petrol engine draws a mixture of petrol and air during suction stroke.
- A diesel engine draws only air during suction stroke.
- The inlet valve is open and exhaust valve is closed during suction stroke.

- During compression stroke, both valves (inlet and exhaust) are closed and the piston moves upward.
- During the expansion or power stroke, both valves remain closed.
- This action of removing out the exhaust gases with the help of fresh charge is known as Scavenging.
- First Law of Thermodynamics can be stated as whenever a system undergoes a cyclic change, the algebraic sum of work transfer is proportional to the algebraic sum of heat transfers or work and heat are mutually convertible one into the other.
- The heating of gas at constant pressure is governed by Charles law.
- The heating of gas at constant temperature is by boyle's law.
- In the equation pv^n= constant, the process is said to be hyperbolic if then is equal to 0.
- If the value of n is equal to zero in the general law, process is said to be isobaric.
- Swept volume or all pistons during power stroke/min is called displacement volume.
- BHP is the horse power available at crank shaft.
- Stroke bore ratio is the ratio of stroke length/ diameter of pistion.
- Number of cams on camshaft is equal to number of tappets.
- Inlet and exhaust valves both will remain open for a moment at the starting of suction stroke.
- In cylinder block piston liners are sealed with the water jacket by O-ring.
- Oil pump is driven by timing gear.
- Differential is compound gear train.
- In two stroke engine transport port opening is exactly opposite and in level with exhaust port.
- Typical value of tappet valve clearance ranges between 0.1-0.4 mm.
- At the end of exhaust stroke both inlet and exhaust valves are seen open for little time.
- When differential lock is engaged in a two wheel tractor, neither axle can transmit more than is permitted by the wheel with poorest traction.

- The most used and least efficient power outlet of tractor is drawbar power.
- The toe in provided in a tractor is approximately 7-10 mm.
- The temperature with in the cylinder of an internal combustion engine of a tractor during fuel combustion ranges between 1650-2200°C.
- In an adiabatic process system the heat is constant.
- Thermostat of petrol engine is fully opened at 82°C.
- Final drive is provided on tractors to increase torque.
- Standard PTO speed recommended by BIS is 540±10 rpm.
- The change of entropy when the heat is absorbed by the gas is positive.
- Valve clearance is adjusted when both the valves are in closed position.
- Fly wheel is made of cast iron.
- Bore is diameter of engine cylinder.
- Weight transfer in a tractor implement system is caused by application of pull.
- White smoke indicates presence of water in fuel.
- Blue smoke indicates burning of lubricating oil in cylinder.
- Cold spark plug is used on heavy engines.
- In diesel cycle the heat is added at constant pressure.
- The traction coefficient is maximum in a field when it is dry.
- Thermal efficiency of a diesel engine is 32-38 %.
- The maximum side thrust is found in roller follower.
- A hole is provided in the crankcase for dipstick in case of four stroke engine.
- Mica can be used as lubricant for heavy engines.
- Inlet valve close at 30° after BDC.
- In case of down daft carburetor, carburetor is mounted above the intake manifold of the engine.
- Thermostat valves are designed to start opening at 70-75°.

- The temperature at which the oil just flows under prescribed conditions is known as pour point.
- Pressure due to radiator cap increases boiling point temperature.
- In a forces feed lubrication system of an IC engine the oil pump is operated by camshaft.
- Exhaust valve of 4 stroke engine opens 45° before BDC and closes at TDC. Thus, the valve remains open for 225°.
- A tractor fitted with diesel engine will have good slogging ability or lugging ability if it has a reserve of 5000 rpm.
- Detonation is favoured by low crank shaft speed at full throttle.
- Total mechanical losses will amount to 10-20 % ihp.
- For optimum working the slip should not exceed 15 %.
- Gears commonly used on fuel injection pumps is rack and pinion type.
- Swinging drawbar is advantageous because it helps taking short turns with machine wider than tractor, reduce side draft, leaves small area uncovered at corners.
- Compression ratio of diesel engine is 15:1.
- Maximum torque in engine is generated at less than rated rpm.
- Tractor operation in a lower gear causes more pull.
- The operation of surface finishing of engine cylinder is called honing.
- If displacement volume is divided by piston displacement no of power strokes per minute for all cylinders will be obtained.
- Enthalpy of a system indicates internal energy of working medium.
- In a tractor the centre of gravity should be located at $1/3^{rd}$ of the wheel base ahead of rear axle.
- The first law of thermodynamics states that all forces of energy are mutually convertible.
- Multiple universal joints are preferred over single joint because they minimize speed fluctuation.
- The hydraulic system used to raise, lower or position an implement either mounted or trailed by moving a hand lever forward or backward from its position is called nudging system.

- Ballasting is done when the wheel slippage exceeds 18% under normal working conditions.
- Tractive efficiency is defined as the ratio of output power to the input power of a traction device.
- Radius of curvature of 60 cm diameter disc with 1cm concavity is 60 cm.
- Water pump of the tractor discharges into the radiator.
- Hydraulic lift capacity upto 65 Kw tractor is about 275 (N/drawbar Kw).
- Standard direction of rotation of crankshaft as viewed from front and numbering of multi cylinder are clockwise & from flywheel end respectively.
- Best speed to get minimum break specific fuel consumption is maximum speed.
- Traction and transmission coefficient for four wheel drive is more than that of two wheel drive tractor.
- Drawbar power output is always less than pto power output due to drive wheel slippage, tractor rolling resistance and frictional losses.
- Clutch in the tractors is designed on the basis of constant pressure.
- Brake testing is done at a speed of 25 kmph.
- Pain in the lower drum starts at a noise level of 115 dB.
- Function of compensating jet in carburetor is to supply lower mixture at higher speed.
- Fuel equivalent power is obtained by multiplying heating value of fuel and rate of fuel consumption.
- Acreman steering consists of sliding pairs.
- Hooks joint is used to connect two parallel offset shafts with short center distance.
- A wheel driven with zero opposing external torque is called driven wheel.
- Hydraulic pump converts hydraulic energy into mechanical energy.
- Thermal efficiency of a tractor in relation to specific energy consumption is inverse.
- Power stroke of a two stroke engine are two times the power strokes of a four stroke engine.

- Chasis dynamometers are required for indoor testing.
- When the stroke bore ratio increased, volumetric efficiency of high speed engine is decreased.
- Air pressure in the front wheels of the tractor is more than that in rear wheels.
- The specific fuel consumption of diesel tractor is lower than that of a petrol engine tractor.
- Turning radius is affected by centre of gravity.
- Specific fuel consumption of diesel engines varies from 250-300 g/Kw-h.
- Volumetric efficiency of tractor varies from 85-90 %.
- As the fuel cutoff approaches to zero, the thermal efficiency of diesel engine approaches that of otto cycle.
- A pressurized radiator cap can help to reduce the evaporation losses by raising the boiling temp of water.
- The use of differential lock in a 4 wheel tractor is to improve the traction by allowing equal torque to be applied to both the rear wheels.
- In high speed diesel engines, the common metal used for piston is aluminum alloy.
- Flow rate of a nozzle is proportional to square root of pressure.
- Specific gravity of high speed diesel is less than that of light diesel oil.
- ASTM- American Society for Testing Materials.
- The ratio of drawbar power to the power input to the wheels axle is tractive efficiency.
- The oil bath type air cleaner uses SAE-40 oil where as tractor gear box uses SAE-90 oil.
- Multi cylinder engine requires lesser weight fly wheel than single cylinder engine.
- The exhaust valve remains open a few degrees after the TDC to improve scavenging.
- A glow plug differs from a spark plug as it helps in cold starting.
- A cetane rating of 66 indicates that a diesel fuel has the same ignition delay in a 60 % n-cetane and 40 % heptane.

- The method of calculating the cetane number was developed by API.
- The lowest temperature at which the fuel ceases to flow is called pour point.
- Liquid petroleum gas commonly known as LPG, consists propane or butane.
- Ethanol has octane number of 111.
- The cloud point will occur 5° C above pour point.
- Higher engine speeds decrease engine life, increase SFC.
- Small one cylinder horizontal diesel engine have been developed widely used for power tiller because of simplicity and lower specific gravity.
- Towed force or motion resistance of a pneumatic tyre is dependent on load size, inflation pressure as well as soil strength.
- In general cone index should be measured before the soil is subjected to wheel traffic.
- Tractor coefficient & tractive efficiency will be greater for 4 wheel drive tractor than that for 2 wheel drive tractor when operating on a soft soil.
- The average weight of a power tiller is in the 200-300 kg range as the average engine power of 7 Kw.
- Higher the thermal efficiency can be obtained with higher compression ratio.
- Higher stroke bore ratio lesser the thermal efficiency.
- The beveled poppet valve is used on valve timing.
- The valve ports are commonly designed to produce a maximum gas velocity of 7 m/s in exhaust port, 56 m/s in in the inlet valve port.
- Piston attains its maximum velocity at 75° to 80° from TDC at which the angle between the crank case and connecting rod is being perpendicular.
- Rocking vibrations in the plane formed by cylinders of an inline engine can exist in two and three cylinder engines.
- Peak pressure is highest for open chamber design.
- Viscosity numbers without an additional symbols are based on the viscosity at 99°C.
- Viscosity numbers with additional symbol are based on viscosity at -18°C.
- Reference sound pressure level 0.00002 N/m^2.

- Sound pressure level doubles with increase in 6 dB.
- The two reference lubricants have same viscosity at 99°C.
- The transition point between the braked & driven force is the towed wheel condition.
- The transition point between drive & driving force states the self-propelled wheel condition.
- Longitudinal stability of tractor may be increased by adding weight to the tractor chasis in such a manner as to move forward and lower the chasis centre of gravity.
- A tractor with a pivoting front axle assembly may tip sideways about two axes.
- Inverting the cornering coefficient slip angle relationships for the front wheel gives the angle.
- Moment of inertia of a tractor about the longitudinal, traverse or vertical axis passing through its centre of gravity.
- A spur gear pump is normally used on tractor hydraulic system of low pressure.
- The spur gear, the internal gear pump the generator gear pump and vane type pump are all used on tractor hydraulic system where lower pressure are used.
- An accumulator is the only convenient device for storing energy in a hydraulic system.
- An accumulator also serve as a shock absorber to reduce maximum stresses if the system is subjected to unusual dynamic loads.
- Bulk modulus is the most important property in determining the dynamic performance of hydraulic systems because it relates to the stiffness of the liquid.
- Dynamic viscocity - Ns/m^2.
- Kinematic viscocity - m^2/s.
- The system used to raise, lower or position an implement either mounted or trailed by moving hand lever either forward or backward from its neutral position.
- In control language the nudging system is an open loop system.

- Draft sensing was located on the upper link & responded to a compressive force.
- Open centre constant flow system normally requires larger diameter valves than closed centre systems to provide for flow loss in the centre.
- The three point hitch was developed in 1935 by the late harry ferguson.
- The quick attaching coupler moves the implement rearwards approximately 10cm.
- Sliding gear is spur gears.
- Constant mesh gears are helical gears, always remain in contact and are less efficient than sliding gears.
- CVT- continuously variable transmission.
- Spur gears have teeth that are parallel to the axis of rotation of the gear where as the teeth of helical gears are at angle to the axis of rotation.
- Bevel gears used to connect shafts with intersecting axes & shafts intersect at right angles.
- Straight bevel gear is the bevel gear equivalent of the spur gear with load transferring abruptly from tooth to tooth.
- Spiral bevel gears have curved teeth that are angled like helical gear teeth so that the load is transferred gradually.
- Zero bevel gear is a special type of spiral bevel with zero spiral angle.
- A hypoid gear is a special type of bevel gear but shafts donot intersect.
- The general range of tooth modules for various gears in the drive train is as:
  - Transmission gears- 4-5
  - Power shift planetary gears- 2.5-3.5
  - Spiral bevel gears 8 to 12
  - Final drive gears 5 to 7.
- The differential assembly normally includes a large spiral bevel gear that drive the differential housing gears.
- The function of a differential lock used in a rear wheel drive tractor is to operate both the rear wheels at same speed.

- Thermal efficiency of a tractor engine is about 33%.
- The volumetric efficiency of a naturally aspirated diesel engine is 85-90%.
- Otto, 1876 invented the first successful internal combustion engine.
- J. froelich is the inventor of the tractor.
- The first tractor company introduced air cooled engine in india was eicher.
- Sound level of tractor for 8 hours operator exposure should not exceed 90 db.
- When ratio of tractor chasis frequency to set natural frequency is more than 2 the viscous damping increases the vibration transmissibility.
- Operator tolerance to vertical vibration for 8 hours duration is 0.315m/s^2 (RMS).
- Size of tyre is normally expressed as tyre section width X outer dia.
- The size of tractor tyre may be represented as section thickness X rim diameter.
- Ply rating of tyre represents the load bearing capacity.
- Oil to be used in engine for lubrication purpose is SAE-30.
- Lubricant used in gear box is SAE-90.
- The grade oil used in winter seasons is SAE-30.
- The grade oil used in summer is SAE-40.
- The voltage in spark plug at the time of spark is 20000V.
- The standard test track for drawbar test should have good adhesion characteristics with a coefficient of traction in the range of 0.6-0.7.
- Compression pressure of DI tractor engine ranges from 262-362Kpa.
- Type of fuel injection nozzle generally used in Indian tractors is multiple hole type.
- In I head engine both valves are provided in the cylinder head.
- Engine main bearing tightening torque is usually kept 24-45 kg-m.
- Injection of fuel in CI engine commences at an angle of 5-10° before TDC.
- Gear generally used in transmission box are helical & spur gears.

- Radial ply tyres can be operated at more inflation pressure than the biased ply tyres.
- Typical values for coefficient of rolling resistance on settled tilled soil are 0.15-0.35.
- Air standard engine cycle efficiency of petrol engine in comparison to diesel engine gor the same compression ratio is higher.
- Most preferred firing order of six cylinder engine is 1-5-3-6-2-4.
- For smoother revolution of crank shaft the best firing order for four stroke, 4 cylinder engine is 1-3-4-2.
- In three cylinder inline engine, the power impulse occurs after every 240° of crank shaft revolution.
- Suggested difference in mass to power ratio between ballasted and unballasted tractor is of the order 6kg/pto Kw.
- Weight per unit of drawbar power is generally ranges between 50-120 kg/Kw.
- Dual wheels have disadvantage of concentrating load over an area.
- Tandem wheels are preferred over dual wheels because they tends to reduce load on an area by 50%.
- Weight transfer from front to rear wheels is advantageous in case of 2 wheel drive tractor.
- The lower value of the ratio of drawbar height to length of tractor is better as variation in weight transfer is decreased.
- When the valves are carried in the engine block at one side of the cylinder is called L head.
- Turbo chargers or super chargers increase the power of the engine by 10-15%.
- Turbo chargers are driven by exhaust gases of the engine.
- Super chargers are driven by direct connection from engine power.
- Chemically correct air fuel ratio of gasoline engine is 15.05:1.
- Recommended engine oil for use in winter is SAE-30.
- Cooling system handles about one third of the heat provided by the fuel.
- Specific gravity of battery electrolyte at fully charged condition is 1.280.

- In direct injection of diesel engines ignition delay is shorter.
- A cylinder block is normally made of cast iron.
- In a two port two stroke engine with crank case compression scavenging will be poor, efficiency will be low, lubrication will be difficult.
- The compression ratio of the diesel engine are in the ratio 14:1 to 20:1.
- Maximum noise level from a tractor near the operator ear should not exceed 85db.
- A connecting rod of a tractor engine is made of forged steel.
- Nozzles of diesel engines operate usually at a pressure of 125-175 kg/cm^2.
- The optimum inclination of star wheel inclination of reaper windrower is 32°.
- The optimum velocity ratio of reaper cutter bar to machine forward speed is 0.75-1.
- The heating of gas at constant pressure is governed by Charles law.
- The heating of gas at constant temperature is by boyle's law.
- In the equation pv^n= constant, the process is said to be hyperbolic if the n is equal to 0.
- If the value of n is equal to zero in the general law, process is said to be isobaric.
- Swept volume or all pistons during power stroke/min is called displacement volume.
- Bhp is the horse power available at crank shaft.
- Stroke bore ratio is the ratio of stroke length/ diameter of pistion.
- Number of cams on camshaft is equal to number of tappets.
- Inlet and exhaust valves both will remain open for a moment at the starting of suction stroke.
- In cylinder block piston liners are sealed with the water jacket by O ring.
- Oil pump is driven by timimg gear.
- Differnential is compound gear train.
- In two stroke engine transport port opening is exactly opposite and in level with exhaust port.

- Typical value of valve tappet clearance ranges between 0.1-0.4 mm.
- At the end of exhaust stroke both inlet and exhaust valves are seen open for little time.
- When differential lock is engaged in a two wheel tractor, neither axle can transmit more than is permitted by the wheel with poorest traction.
- The most used and least efficient power outlet of tractor is drawbar power
- The toe in provided in a tractor is approximately 7-10 mm.
- The temperature with in the cylinder of an internal combustion engine of a tractor during fuel combustion ranges between 1650-2200°C.
- In an adiabatic process system the heat is constant.
- Thermostat of petrol engine is fully opened at 82°C.
- Final drive is provided on tractors to increase torque
- Standard PTO speed recommended by BIS is 540±10 rpm and 1000± 50 rpm
- The change of entropy when the heat is absorbed by the gas is positive.
- Valve clearance is adjusted when both the valves are in closed position.
- Flywheel is made of cast iron.
- Bore is diameter of engine cylinder.
- Weight transfer in a tractor implement system is caused by application of pull.
- White smoke indicates presence of water in fuel.
- Blue smoke indicates burning of lubricating oil in cylinder.
- Cold spark plug is used on heavy engines.
- In diesel cycle the heat is added at constant pressure.
- The traction coefficient is maximum in a field when it is dry.
- Thermal efficiency of a diesel engine is 32-38%.
- The maximum side thrust is found in roller follower.
- Conversion efficiency of solar energy into electrical energy is 04-10%.
- A hole is provided in the crankcase for dipstick in case of four stroke engine.
- Mica can be used as lubricant for heavy engines.

- Cam and follower includes higher pair.
- Inlet valve close at 30° after BDC.
- Drawbar power is the product of ground speed and drawbar pull.
- Track laying vehicles are characterized by high drawbar pull and low speeds.
- Two-wheel drive tractors are designed for lighter pulls and faster speeds than track layers.
- For similar tractive conditions a two-wheel drive tractor will achieve a lower pull ratio than a track because of less lug-soil interaction and higher contact pressures.
- The efficiency of transmitting axle power to useful drawbar power is also lower.
- Pull ratio increases continuously with slip while tractive efficiency decreases at higher slip values.
- The Central Farm Machinery Training and Testing Institute, Budni is the only authorized Institute of Govt. of India for testing large Agricultural machinery viz. Tractors, Combine Harvesters, Power Tillers etc.
- Regional Testing Institutes' functioning at Hisar (Haryana), Garladinne (Anantpur) and Biswanath Chariali (Assam), undertake testing of machines below 10 HP and implements.
- The performance of parking brakes is checked by parking the tractors on 12% and 18% slopes.
- The brake performance is checked under cold brake and hot brake conditions at 25 kmph speed.
- As per ASABE standards three point hitch of a two wheel drive tractor with maximum drawbar of 45Kw comes under category II.
- The longitudinal stability of a rear wheel driven tractors moving up a slope in hilly areas for haulage work can be improved by adding weight.
- The rear furrow wheel in a tractor mounted disc plough is provided to reduce the side draft.
- Type of pump used in forced water cooling system of a tractor engine is centrifugal pump.

9

Erogonomics in Agriculture

Ergonomics (also known as Human Engineering, Human Factors or Human Ergology) is the scientific study of relationship between a person and his/her working environment. The term environment includes the tools, materials and method of work, ambient conditions and physical environment in which the work is carried out, and also the organizational factors. Ergonomics application can help in increasing the efficiency and productivity, reducing drudgery and occupational health problems, and enhancing safety of workers in various activities.

- Ergonomics (also known as Human Engineering, Human Factors or Human Ergology) is the scientific study of relationship between a person and his/her working environment.
- The performance of tool/ equipment depends not only on the constructional features but also on the workers operating it.
- While designing any equipment the following aspects should be given importance: functional, structural, economic, ergonomic, and performance aspects.
- The branch of ergonomics which deals with body measurements is called as anthropometry.
- The anthropometric data are used to determine the size and shape of handles, height of work surfaces and the space in which an operator has to work.
- There are 16 strength parameters important in agricultural machinery design.
- Anthropometric criteria fall into four main categories and deal with issues of Clearance, ii) Reach, iii) Posture and iv) Strength.
- Clearance criterion deals with concern like head room, leg room and so on.
- Reach includes those concerned with the location of controls or the storage of materials, and with a variety of situation where it is necessary to reach to perform a task.

- Posture includes those concerned with the location of displays and controls at the height of working surfaces.
- Strength is applicable where a worker has to apply force to the work.
- Vital capacity is the maximum amount of air a person can expel from the lungs after a maximum inhalation.

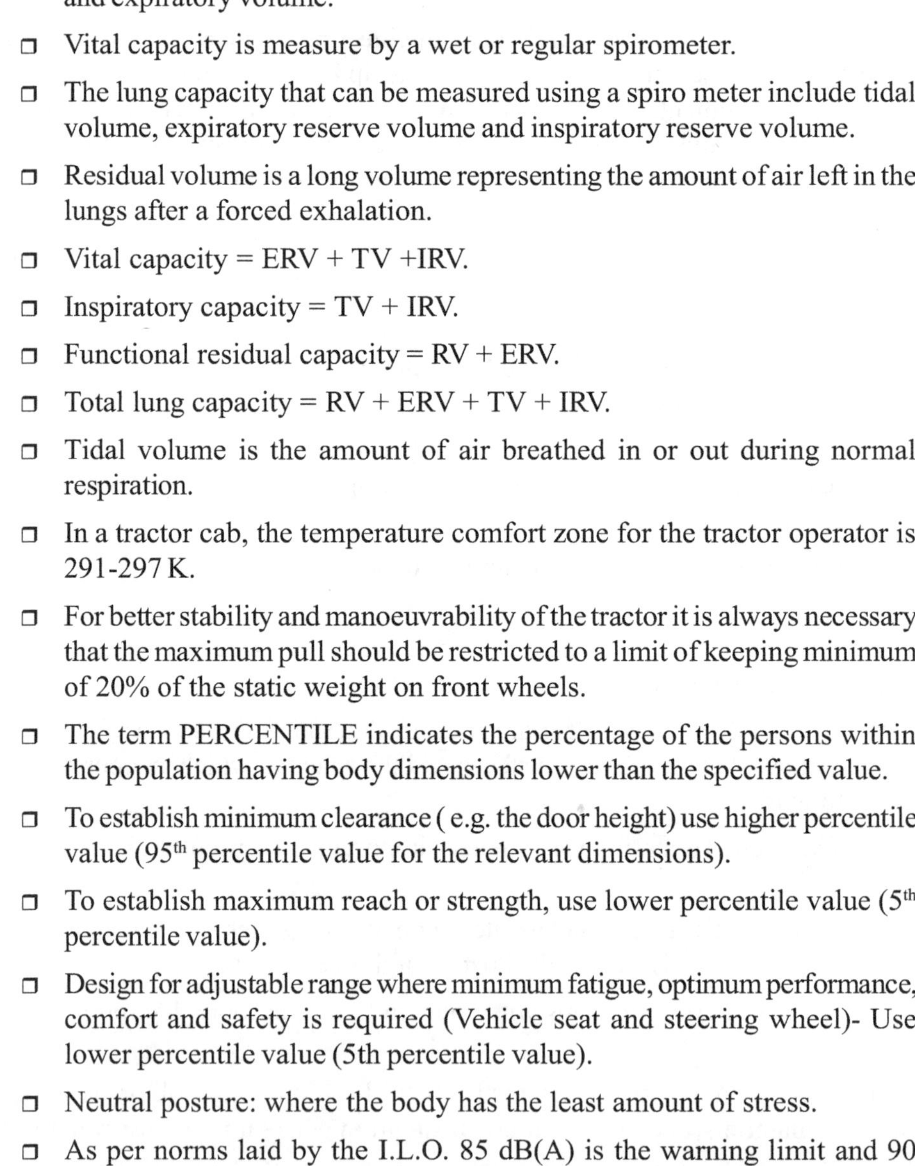

- Vital capacity is equal to sum of inspiratory reserve volume, tidal volume and expiratory volume.
- Vital capacity is measure by a wet or regular spirometer.
- The lung capacity that can be measured using a spiro meter include tidal volume, expiratory reserve volume and inspiratory reserve volume.
- Residual volume is a long volume representing the amount of air left in the lungs after a forced exhalation.
- Vital capacity = ERV + TV +IRV.
- Inspiratory capacity = TV + IRV.
- Functional residual capacity = RV + ERV.
- Total lung capacity = RV + ERV + TV + IRV.
- Tidal volume is the amount of air breathed in or out during normal respiration.
- In a tractor cab, the temperature comfort zone for the tractor operator is 291-297 K.
- For better stability and manoeuvrability of the tractor it is always necessary that the maximum pull should be restricted to a limit of keeping minimum of 20% of the static weight on front wheels.
- The term PERCENTILE indicates the percentage of the persons within the population having body dimensions lower than the specified value.
- To establish minimum clearance (e.g. the door height) use higher percentile value (95th percentile value for the relevant dimensions).
- To establish maximum reach or strength, use lower percentile value (5th percentile value).
- Design for adjustable range where minimum fatigue, optimum performance, comfort and safety is required (Vehicle seat and steering wheel)- Use lower percentile value (5th percentile value).
- Neutral posture: where the body has the least amount of stress.
- As per norms laid by the I.L.O. 85 dB(A) is the warning limit and 90 dB(A) is the danger limit for an continuous exposure of 8 hours.

- The average amplitude of mechanical vibration at the seat, foot rests & steering wheel should not exceed 100 microns for better operational comfort.
- The oil pullover to the engine should not be more than 0.25%.
- The smoke level measured at 80% of maximum power should not exceed light absorption co-efficient of 3.l/m (5.2 Bosch smoke units or 75 Hartridge smoke units).
- Physiological cost of any activity may be expressed in terms of cardio-respiratory responses of the workers and the main parameters used here are heart rate (HR) and oxygen consumption rate (VO_2).
- For calculation of energy expenditure of Indian workers, the calorific value of oxygen can be taken as 20.9 kJ l^{-1}.
- The concept of "Acceptable Work Load" (AWL), which depends on the maximum oxygen uptake or maximum aerobic power (VO_{2max}).
- According to Saha et al. (1979), the acceptable workload for average young Indian worker is about 35% of VO_{2max}.
- For Indian workers the VO_{2max} values can be taken as about 2.2 l min^{-1} for male workers and 1.6 l min^{-1} for female workers, which corresponds to an energy expenditure rate of 46.0 and 33.4 kJ min^{-1}, respectively.
- The postural stress gets expressed in the form of muscular discomfort.
- Muscular discomfort assessed through electromyography (EMG) or by using psychophysical rating scales such as Borg (1970) scale, 0-10 Visual Analogue Discomfort scale or Corlett and Bishop (1976) scale.
- The mean stature (height) and weight of Indian female agricultural workers are 151.5 cm and 46.3 kg as against 163.3 cm and 54.7 kg for male workers.
- The operator work place of a machine is designed relative to seat index point (SIP) and seat reference point (SRP).
- The steering column angle (a) shall be in the range of 0-40°.
- The 5th percentile value of popliteal height sitting of male agricultural workers is 367 mm.
- Four bar linkage type suspension mechanism is generally selected for the tractor seat.

10

Energy in Agriculture

Energy is defined as the capacity of doing work and is necessary for survival. It is basic input to any society's development and economic growth. Everything happens in the world is the expression of flow of energy from one form to other. There are different sources of energy, i.e., chemical, mechanical, fossil, nuclear and renewable energy.

The following products are obtained from refining of crude oil

i) Petroleum gas (below 40°C) used as Liquified Petroleum gas (LPG).

ii) Petrol/gasoline (40 - 170°C) for use in light vehicles.

iii) Kerosene (150 - 250°C) for household and industrial uses.

iv) Diesel (150 - 350°C) for use in heavy vehicles.

v) Residual oils (lubrication oils, paraffin wax and asphalt).

vi) Fuel oil (350 - 400°C) for use in boilers and furnaces.

Biomass is organic material and contains stored energy from the sun. Plants absorb the sun's energy in a process called photosynthesis. The periodic rise and fall of water level of sea, which is carried by the action of the sun and moon on water of the earth is called "tide".

- Solar energy can be converted into thermal and electrical forms of energy.
- India's wind energy potential is 45000 MW.
- Heat is used in the conversion of biomass into various end use energy forms in thermochemical energy conversion route.
- Biochemical conversion technologies utilize microorganisms during the conversion of biomass to end use forms of energy.
- Trans esterification process which converts vegetable oils to biodiesel.
- Degree of saturation of air is given in terms of relative humidity.
- Ratio between active power and apparent power is given as power factor.

- The measure of energy content of the material is termed as calorific value.
- Octane number is a value used to indicate the resistance of a motor fuel to knock.
- Biological oxygen demand is the indicator of the presence of organic matter in the given sample.
- Elemental analysis of samples may be carried out using Elemental Analyser.
- Glycerol liquid is used in the determination of density of hygroscopic nature materials.
- Kjeldahl method is used to determine the nitrogen content.
- Calorific value can be calculated based on elemental composition of biomass using Dulong's formula.
- The temperature for the assessment of charcoal yield from biomass is 200-350°C.
- The molecular weight of carbon dioxide is 12.
- 4 kg of oxygen is required to burn one kg of methane.
- The amount of oxygen in air is 23% by weight.
- Biodiesel is obtained by the process of transesterification.
- Producer gas is obtained by the process of pyrolysis.
- Ethanol is obtained by the process of yeast fermentation.
- Biogas is obtained by the process of anaerobic digestion.
- If the fuel has one kg of sulphur, on complete combustion it produces 2 kg of SO_2.
- More amount of oxygen in the combustion product indicate inefficienct combustion.
- The thermal efficiency of conventional wood stoves are about 8–10%.
- The reason for higher thermal efficiency of improved wood stoves are better air supply and reduced heat loss.
- The thermal efficiency of TNAU single pot chulha is 20–25%.
- The type of gasifier which produces nearly tar free gas is co-current gasifier.

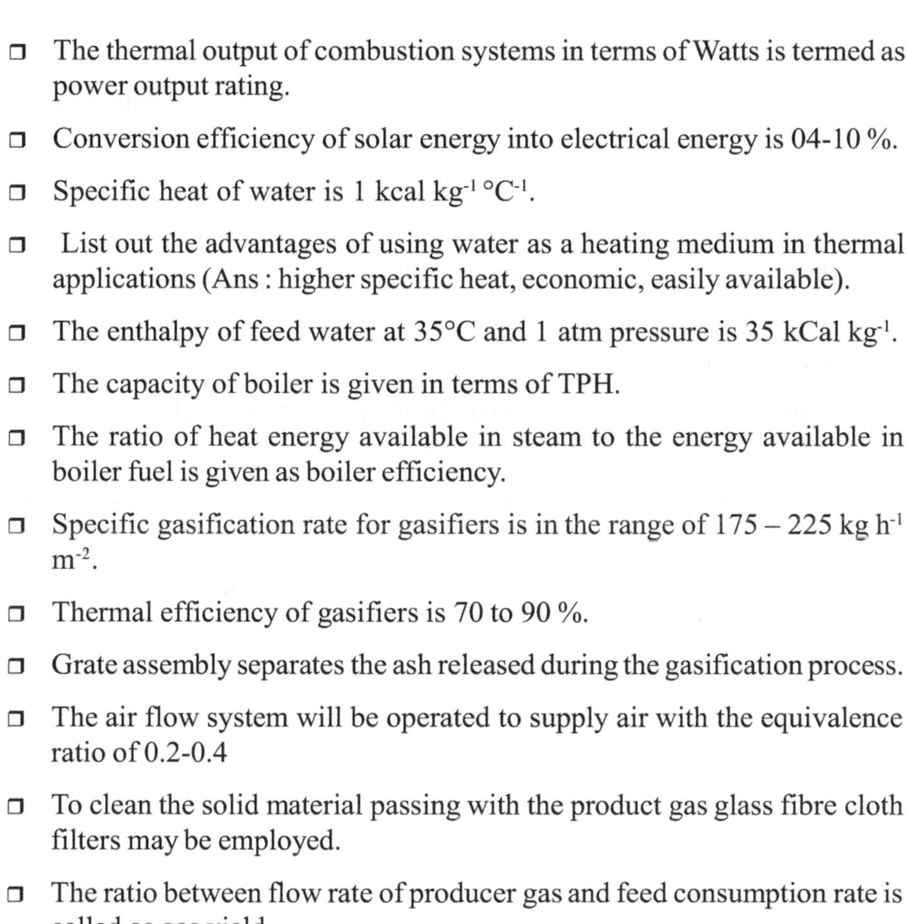

- The thermal output of combustion systems in terms of Watts is termed as power output rating.
- Conversion efficiency of solar energy into electrical energy is 04-10 %.
- Specific heat of water is 1 kcal kg^{-1} $°C^{-1}$.
- List out the advantages of using water as a heating medium in thermal applications (Ans : higher specific heat, economic, easily available).
- The enthalpy of feed water at 35°C and 1 atm pressure is 35 kCal kg^{-1}.
- The capacity of boiler is given in terms of TPH.
- The ratio of heat energy available in steam to the energy available in boiler fuel is given as boiler efficiency.
- Specific gasification rate for gasifiers is in the range of 175 – 225 kg h^{-1} m^{-2}.
- Thermal efficiency of gasifiers is 70 to 90 %.
- Grate assembly separates the ash released during the gasification process.
- The air flow system will be operated to supply air with the equivalence ratio of 0.2-0.4
- To clean the solid material passing with the product gas glass fibre cloth filters may be employed.
- The ratio between flow rate of producer gas and feed consumption rate is called as gas yield.
- Tar, char, dust and carryover particles are to be removed from producer gas before supplying to the engine applications.
- Orifice meter instrument is used to measure the flow of air to the gasifier system.
- Define specific gas production rate. (Ans : The ratio between weight of feed material used to cross sectional area of the reactor).
- Hot gas efficiency is the ratio between output energy (producer gas) and input energy (fuel).
- The maximum temperature of gasifier reactor during gasification is about 1200 °C.
- Tuyers are the air distribution system, having close contact with the oxidation zone of gasifier.

- Coarse particles are separated from the gas stream in a cyclone separator.
- List out various daily maintenance of the gasifier system. (Ans : Removal of ash, unburnt fuel and tar condensate).
- The floating drum of KVIC biogas plant is made up of MS or FRP.
- The biogas plant has to be fed with cowdung with water in a mix ratio of 1:1 to maintain the optimum total solid content.
- The entire deenbandhu biogas plant should be covered with soil. (True / False) (Ans : True).
- The flammability of biogas can be tested at the gas outlet of the biogas plant. (True / False) (Ans : False).
- For every kg of destruction of COD, about 0.3 m^3 of methane can be generated.
- Ferroin indicator is used in the assessment of Biological Oxygen Demand using chemical method.
- The gas calorimeter can be used to assess the calorific value of biogas.
- The saccharometer is used to measure the methane content of biogas sample.
- The carbon dioxide present in the biogas will be dissolved in NaOH solution in saccharometer.
- For every kg of destruction of COD, about 0.3 m^3 of methane can be generated.
- Ferroin indicator is used in the assessment of Biological Oxygen Demand using chemical method.
- The gas calorimeter can be used to assess the calorific value of biogas.
- The saccharometer is used to measure the methane content of biogas sample.
- The carbon dioxide present in the biogas will be dissolved in NaOH solution in saccharometer.
- In KVIC model, the gas holder is given half rotation 2-3 times to break the scum on the surface of the digester slurry.
- About 25 kg of cattle dung is required for generation of 1 m^3 gas per day.
- Soap solution may be used for testing leakage of gas in the dome/gas holder. (True / False) (Ans : True).

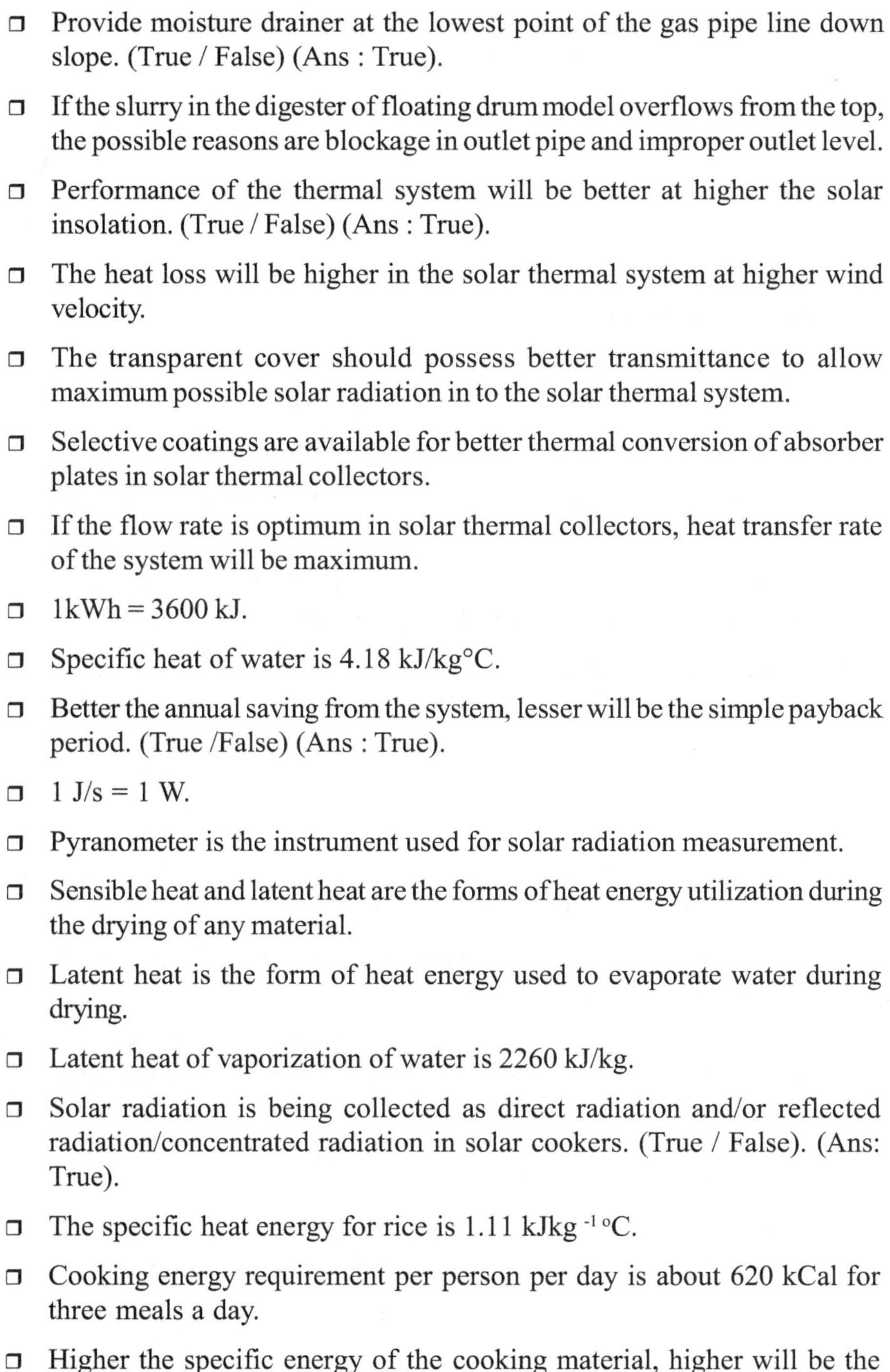

- Provide moisture drainer at the lowest point of the gas pipe line down slope. (True / False) (Ans : True).
- If the slurry in the digester of floating drum model overflows from the top, the possible reasons are blockage in outlet pipe and improper outlet level.
- Performance of the thermal system will be better at higher the solar insolation. (True / False) (Ans : True).
- The heat loss will be higher in the solar thermal system at higher wind velocity.
- The transparent cover should possess better transmittance to allow maximum possible solar radiation in to the solar thermal system.
- Selective coatings are available for better thermal conversion of absorber plates in solar thermal collectors.
- If the flow rate is optimum in solar thermal collectors, heat transfer rate of the system will be maximum.
- 1kWh = 3600 kJ.
- Specific heat of water is 4.18 kJ/kg°C.
- Better the annual saving from the system, lesser will be the simple payback period. (True /False) (Ans : True).
- 1 J/s = 1 W.
- Pyranometer is the instrument used for solar radiation measurement.
- Sensible heat and latent heat are the forms of heat energy utilization during the drying of any material.
- Latent heat is the form of heat energy used to evaporate water during drying.
- Latent heat of vaporization of water is 2260 kJ/kg.
- Solar radiation is being collected as direct radiation and/or reflected radiation/concentrated radiation in solar cookers. (True / False). (Ans: True).
- The specific heat energy for rice is 1.11 $kJkg^{-1}\,{}^{o}C$.
- Cooking energy requirement per person per day is about 620 kCal for three meals a day.
- Higher the specific energy of the cooking material, higher will be the energy required for cooking. (True / False) (Ans : True).

- The cleaned transparent cover will give better transmittance of solar radiation. (True / False) (Ans : True).
- Scaling in the flow pipes will reduce heat transfer from solar collector to the medium.
- Damages in the insulation will cause more heat loss from the collector area.
- Selective coatings gives better absorptance of transmitted solar radiation into the solar collectors.
- Optimum mass flow rate of heating medium will lead to better heat transfer and better performance of solar heating devices.
- Individual solar modules are connected to form arrays.
- The reliability of solar photovoltaic systems is higher as they are not having moving parts.
- Combining PV systems with any type of power production systems such as wind, hydro, diesel is called as hybrid system.
- In the performance assessment of solar street light system, the system efficiency includes panel efficiency and solar street light system efficiencies.
- Total head to be pumped is the addition of suction, delivery and friction heads.
- Do not work with the system while connected with the SPV or with working condition. (True /False) (Ans : True).
- Ensure working of charge controllers to maintain the battery charging and discharging cycle.
- Major check-ups in battery are checking of specific gravity and temperature in each cell in battery. Based on application of pressure, briquetting production of the agro residues are divided into three categories; they are high pressure or high compaction technology, medium pressure technology and low pressure technology.
- Two major types of binder less or high compaction briquetting technologies are piston press (ram and die) technology and screw press technology.
- Most of the briquetting machines installed in India are piston press type.
- The power consumption in piston press technology is lesser than screw press briquetting technology.

- The performance of the drying system will be assessed in terms of efficiency of the system, capacity and quality of the final product.
- Lesser the size of the feed material better will be the briquetting of crop residues. (True / False) (Ans : True).
- The raw materials are reduced to uniform size of about 1-10 mm before briquetting.
- The briquetting system efficiency is the ratio between the input power supplied to the system to the energy spent for the production of briquettes in a specified time.
- The quality of the briquetting product relay on the moisture content of the feed crop residues. (True / False) (Ans : True).
- The proximate components of the feed materials are volatile matter, fixed carbon and ash content. Bulk density of biomass briquettes is found by observing the weight of known volume of briquette.
- The efficiency of a pair of bullocks is less than efficiency of single bullock in field.
- Calorific value of biogas is about 3400 Kcal/m^3.
- To assess the water resistance briquette is immersed in water at 27 °C for 1 min.
- Briquettes may be dropped from about 1.5 m height onto a metal plate for 4-5 times to assess the impact resistance / drop resistance / shattering resistance property.
- The water dipped briquettes which are stable for more than 15 min is considered as good quality briquettes.
- The common three types of balers are square balers, large square balers and round balers.
- Round bales are effective for large scale operations with minimal cost.
- The basic modes of operation of balers during preparation of bales are loading, compression and ejection.
- The problems of higher moisture content of raw materials lead to poor core formation and blockage problems in baling.
- After ejection, if the large (weighing about 500 kg) balers are starting to roll, try to stop to avoid further damage. (True / False) (Ans : False).

- Direct substitution of triglycerides in diesel engines will lead to the engine problems. (True / False) (Ans : True).
- The fatty acid alkyl esters from the trans esterification reaction are called as biodiesel.
- Reaction temperature for biodiesel production is 60 to 70 °C.
- The optimized dosage of alkaline catalyst concentration is in the range of 0.5 to 1 % by weight.
- The reaction temperature of the transesterification process will be decided based on the temperature of vaporization of selected alcohol.
- The higher mixing speed may cause the disturbance of the reactants resulted to hindering the reaction. (True / False) (Ans : True).
- Theoretically the addition of alcohol is in 3:1 molar ratio to the moles of vegetable oil.
- The problem of higher addition of alcohol is difficult in segregation of esters.
- NaOH and KOH are the most common catalyst used in Trans esterification process.
- Biodiesel has the problem of gelling during storing at low temperatures. (True / False) (Ans : True).
- Oxidative stability is the ability of the biodiesel to be stored for extended periods without oxidation.
- Generally the acceptable materials for biodiesel storage are aluminium, steel , Teflon.
- There is no lead or sulfur to react and release any harmful or corrosive gases from biodiesel. (True / False) (Ans : True).
- B40 means 40 % of biodiesel and 60 % of diesel mixture.
- Variable compression ratio multi-fuel engine test rigs have facilitates of compression ratio from 5:1 to 20:1.
- Exhaust analyser or multi gas analyser may be used to study the exhaust emission during engine testing experiments.
- Specific gravity is the ratio of the density of a substance to the density of a reference substance.

- Cetane number is a measure of the quality of a diesel fuel expressed as the percentage of cetane in a mixture of cetane and 1-methylnapthalene of the same quality as the given fuel.
- Viscosity is a measure of resistance to flow of a liquid created due to the internal friction of one part of a fluid moving over another.
- The pour point of a liquid is the lowest temperature at which it becomes semi solid and loses its flow characteristics.
- Acid number reflects the presence of free fatty acids or acids used in the manufacturing of biodiesel.
- Brick kilns has a long production cycle.
- A pyranometer is used to measure indirect solar radiation.
- In updraft gasifier the height to diameter ratio is kept 3:1
- Charcoal yields from pits is very low.
- In fluidized bed reactor fluid is pyrolyzed very fast.
- Moisture influence the temperature in various zones of gasifiers.
- Hard burned charcoal depends on its friable properties.
- Quality of charcoal depends on its physical and chemical properties.
- Ethanol is a primary product.
- Diesel fuel is only used for starting dual fuel engine.
- Petrol engine can be run on 100% biogas.
- Charcoal yields from cement kilns can be high as 25-33 %.
- Sugar cane is a common feed stock for ethanol.
- Species of timber can be used for alcohol production.
- In fixed dome type biogas unit the gas is liberated at variable pressure.
- In fixed dome type biogas unit diameter and height ratio of digester fixed at 1.75:1.
- In fixed dome type biogas unit the volume of dome is 60% of the total contents.
- Water content in biogas plant slurry should be about 90% of the weight of the total content.

- About 70-75 % of the original weight of cattle dung is conserved in a biogas unit.
- Biogas dual fuel engine requires 15-20% of diesel and rest of the energy is obtained from biogas.
- A gap of 50 mm is normal between the gas forming on the sides of plant into gas holder.
- The maximum theoretical efficiency of the horizontal axis type of machine approximately 60 %.
- Convertible wind power or energy is proportional to cube of the wind speed.
- The savonious rotor requires relatively low velocity winds for operation.
- The Darrenious machine is a type of vertical machine.
- Nonrenewable energy is energy obtained from static stores of energy that remains bound unless released by human interaction.
- Nuclear fuels, fossil fuels of coal, oil and natural gas are examples of non-renewable energy sources.
- Solar pond used to collect and store the solar energy.
- High wind energy region in India is Gujarat.
- A sail boat utilize the energy of wind.
- The major constituent of biogas is methane.
- Digester diameter of biogas ranges between 1.2-6 m.
- In biogas plants, the depth of digester varies from 3-6 m.
- The percentage of CO_2 in methane is 32-45 %.
- The economically smallest size of gobar gas plant is $70 ft^3$.
- KVIC biogas plant is floating type.
- In fixed dome type biogas plant the gas holder is made of cement/briks.
- In floating type biogas plant the gas holder is made of metal.
- A high gas production achieved from the biogas plants when the inside temperature of chamber is 35°C.
- The temperature inside the solar cooker ranges from 100-140°C.

- An instrument used to measure the solar radiation is pyrheliometer.
- An instrument which measures the total or global radiation over a hemi spherical field of view is pyranometer.
- An instrument used to measure wind speed is anemometer.
- The optimum C:N ratio for maximum microbial activity is 30:1.
- According to NASA, the standard value for solar constant is 1.353 Kw/ m^2.
- Fuel cell or electro chemical cell or hydrogen cell has an efficiency of 38-60 %.
- Seebeck effect- if two dissimilar materials are joined to form a loop an d the two joints are maintained at different temperatures, an emf will be generated in the loop.
- Pyrheliometer is for measuring beam radiation.
- Pyranometer is for measuring global or total radiation.
- 60 cm X 60 cm X 20 cm height are the dimensions of typical solar cooker.
- Solidity ratio is the ratio of projectional area of rotor to the swept area of the rotor.
- Total solidity ratio increases the no of blades decreases.
- The gas produced in the gasifier is a clean burning fuel having heating value of about 950-1200 Kcal/m^3.
- Downdraft gasifier is also called concurrent gasifier.
- Methanol is a clear, colorless liquid that freezes at -73°C and boils at 65°C.
- Solar energy absorption by water takes place according to lambert's law of absorption.
- Open cycle of OTEC system is Clark cycle.
- Closed cycle of OTEC system is Anderson cycle.
- Fuel cells are particularly suited for low voltage high current applications.
- In anaerobic digestion biogas production by bacterial decomposition takes place in the absence of oxygen.
- Boiling temperature of methanol is 65 °C
- Paraboidal type solar cooker was first developed in India.

- The gas generated through biomass gasification is called producer gas.
- KVIC type of biogas plant was developed in India.
- Maximum efficiency of solar photo voltaic cell attainable today is 12-15%.
- A typical insulation material for liquid solar collector is glass wool.
- The storage system in electric vehicle for fuel supply is Nickel cadmium rechargeable battery.
- Metal hydride is used for storage of geothermal energy.
- The maximum temperature that can be obtained in a flat plate box type solar cooker ranges between 50-75°C.
- In a box type solar cooker the solar radiation entering or leaving the box are of short/ long wave length respectively.
- Presence of $H\text{-}_{2\text{-}}S$ in biogas causes corrosive effect.
- The efficiency of biogas burner used for family type biogas plant is about 55 %.
- In a diesel engine the ratio of biogas : diesel consumption is about 80:20.
- Wind mills with low tip sped ratio have many blades.
- Wind speed varies with height.
- 1 kg of dung produces 0.04 m^3 /day.
- Retention time for production of biogas is less in summer than in winter.

11

Workshop Practice and Technology

For development of agricultural engineering machineries, materials and manufacturing techniques play a very important role. Functionality, efficacy, life span, economics and aesthetic of any machine or structure are affected severely by the material of fabrication process. The manufacturing processes discussed in this chapter also include the pretreatment to manufacturing and structural materials. Wood and some kind of ferrous alloys had been the choice of materials for fabrication of agricultural machines in past few centuries, whereas these materials along with mud, bamboo, some agricultural residues and some earthen materials were the materials of choice for agricultural structures. Engineering materials are available with wide range of properties and characteristics. There are many properties, which are inherent in the materials, and some of them can be changed during processing and manufacturing.

- Carbon content in cast iron is 2.5-4.5 %.
- Carbon content in mild steel is 0-0.25 %.
- Medium carbon steel has carbon content of 0.25-0.50 %.
- Carbon content of high carbon steel is 0.5-1.5 %.
- Brass is an alloy of copper & zinc having its relative percentage 60-70 % cu & 30-40 % zn.
- Bronze is an alloy of copper, tin & zinc having its relative percentage 88 % cu,10 % tin & 2 % zn.
- Main function of cutout relay in electrical circuit is to prevent the reversal of current flow from the battery when the dynamo speed reduces.
- The tool which is given reciprocating motion by hand is called saws.
- The operation of producing thin-walled hallow or vessel shaped parts from sheet metal is called as drawing.
- The process of cutting unwanted metal with a cold chisel and a hammer is called chipping.

- The tool used for hallow sinking of Head bolt and screws is allen key.
- The means of cutting a whole piece from sheet metal just enough scrap is left all around the opening is called blanking.
- Engineering parallels are made with cast iron/steel.
- The ability of a metal to be hammered, rolled, or pressed into various shapes without rupture or fracture is called malleability.
- The vice that is generally used in plumbing work for holding the hallow circular sections such as pipes and round bars is pipe vice.
- The Screw driver used for turning screws in confined space is rachet screw driver.
- The radial rays or thin layers running between the pith and cambium layer of the exogenous tree are medullary rays.
- The saw used for cutting curves of small radius and cutting excess wood in making dove tail joints is coping saw.
- The tool used for holding and driving the thread die is die stock.
- Cast iron contains >2 Percent of carbon alloy.
- The hand shear is also called as snip.
- The part that supports the cutting edge and also breaks the shavings so that they curl away from the blade in a wooden jack plane is back iron.
- Tap drill size for commercial purpose $D_t = D - P$.
- Mitre square is used for marking and testing of 45°.
- The narrow and thin part of the file which fits into the wooden handle is tang.
- The tools used for necking down a piece of work are Fullers.
- The gauge used for checking the dimension of hole is Plug gauge.
- The chisel used for cutting V-grooves and sharp corners is Diamond point chisel.
- Warding file is used for filing Narrow slots.
- Joint used for widening the Planck length is Groove joint.
- The pitch of the bastard file is 1.6 – 0.65 mm.

- Ingot iron contains 99.85 percentage of Iron.
- The Material allowance left in the hole for hand reaming is usually 0.05 to 0.1 mm.
- The pith of the teeth of crosscut saw are about 2.5 to 3 mm.
- The size of the reamer can be given by its Diameter.
- The ability of a metal to resist being fractured by opposing forces not acting in a straight line Shear strength.
- Block plane is set at an angle of 20° to the sole of the plane with the bevel up.
- Gouges can be referred as Round chisels.
- Chisel used in making mortise joint is Mortise Chisel.
- The hammer used for pulling out the nail is called Claw hammer.
- Accuracy of stick micrometre is ±0.005 mm.
- The process of cutting a recess along the edge of wood is called Recessing.
- Shash cramp is a Holding tool.
- The welding in which the metal parts to be joined are heated to a plastic state over a limited area by their resistance to the flow of an electric current is Resistance welding.
- In fullering operation the axis of the job is positioned perpendicular to the width of the flat die.
- Bridle joint is not a Halving joint.
- Permanent moulds are made with metals.
- Shaper ram have reciprocating Motion.
- Operation of cutting off the unwanted parts, cleaning and finishing the casting is known as fitting.
- Screw cutting lathe was developed in the year 1797.
- Type and maintaining of tool post depends upon class of work.
- The process of producing cavity or Mould is called moulding.
- A replica of desired casting is called pattern.
- The most common type of drill used in drilling machine is twist drill.

- The method which is usually employed for small castings having light weight is known as bench moulding.
- The most commonly used Universal milling machine in general work shop is column and knee type.
- The Movement of the tool in lathe machine relative to the work is termed as feed.
- Boring is the operation of enlarging and truing a hole produced by drilling, punching, casting or forging.
- Milling is the operation of removing metal by feeding the work against a rotating cutter having multiple cutting edges.
- Steady rest in lathe machine having three jaws.
- The Working tool in the drilling machine is removed by pressing drift or tapered wedge in to the slotted hole of the spindle.
- The drilling machine which is designed for drilling small holes at high speed in light jobs is sensitive drilling machine.
- The operation which is used for enlarging the previous drilled hole is known as reaming.
- In Universal milling machine the table can be swiveled to any angle up to 45 degrees on either side of the normal position.
- The process of producing Conical surface on a lathe is called as taper turning.
- The operation of cutting off the unwanted parts, cleaning and finishing the casting is known as Fettling.
- Base of shaper is made with cast iron.
- Feed rate of reaming is about 0.5 to 2 mm/rev.
- The name of the top box in moulding flask is Cope.
- The name of the bottom box in moulding flask is Drag.
- The name of the middle box in moulding flask is cheek.
- Rammer is used for packing the sand around the pattern.
- Pressure test applied to casting which are to be used for containing or carrying gases or liquids.
- The moving part in the shaper machine is Tool.

- For general purpose work, the included angle of the lathe centre is 60°.
- The device used for holding and rotating a piece of work in a lathe is Chuck.
- The device used for holding and rotating a hallow piece of work that has been previously drilled or bored is known as Mandrel.
- Milling operation can performed in a lathe by using special attachments.
- The operation which is used for producing a conical surface from cylindrical work piece is Taper turning.
- Tapping is operation of cutting internal threads of small diameter.
- Forged tools are manufactured from high carbon steel and high speed steel.
- The position of ram relative to work can be adjusted by hand wheel.
- V-Block is used for holding round rods in shaper.
- Cutting angle of flat drill is varies between 90 to 120°.
- The type of holding device in drilling machine used for mass production is Drill jags.
- The Drilling machine suitable for drilling medium to large and heavy work pieces is Radial drilling machine.
- End mill is most widely used milling cutters in profile milling works.
- In a shaper tool the amount of side clearance angle is only 2 to 3°.
- V-blocks is the common example of angular milling.
- Metal contraction takes place in the form of liquid contraction, solidifying contraction and Solid contraction.
- Joseph R Brown invented first universal milling machine (1861).
- Dead cache of lathe is subjected to wear due to Friction.
- Operation of beveling the extreme end of a work piece is termed as Chamfering.
- Process of embossing a diamond shaped pattern on the surface of a work piece is known as knurling.
- Process of turning a converse, concave or of any irregular shape is known as Forming.

- Operation of cutting a work piece after it has been machined to the desired size and shape is called parting off.
- Most common type of sharper is Crank shaper.
- Vertical shapes which are specially designed for machining internal keywords are called key setters.
- Size of shaper is expressed as Maximum length of stroke.
- Cutting time to return time ratio of crank and slotted link mechanism is about 2:1.
- Quick method of holding and locating relatively small and regular shaped work piece is done with vise.
- Angle plate of sharper is made with cast iron.
- Super sensitive drilling machines speed is about 20000 rpm.
- Draft allowance for external surfaces varies between 10 to 25 mm.
- Lathe headstock spindle is made of Carbon & nickel – chrome steel.
- Following mechanism makes the carriage to engage or disengage with lead screw half – nut mechanism.
- Included angle of lathe centre for heavy work is about 75°.
- Feed rate of finish turning is about 0.1 to 0.3 mm.
- Shaper which has only two movements, vertical and horizontal to give the feed is standard shaper.
- Lathe speed of polishing is about 1500 to 1800 m/min.
- In a shaper tool the amount of side clearance angle for cast iron is 2° to 3°.
- The drilling machine in which the operator senses the cutting action at any instant, is called sensitive drilling machine.
- In a 250 mm shaper, the length of stroke may be adjusted from 0 to 250 mm.
- Feed rate of machining vertical surface of shaper is about 0.25 mm/stroke.
- Maximum size of drill of portable drilling machine is 12 to 18 mm.
- Maximum size of holes that the pillar drilling machine can drill not more than 50 mm.
- Plain slot clamp of drilling machine is made with mild steel flat.

- Device used for holding round work pieces is v-blocks.
- Quick change chuck is also known as magic chucks.
- Available sizes of shaper is ranging between 175 to 900 mm.
- A special attachment used for cutting equally special grooves or splines on the periphery of a round work is shaper centres.
- Shaper cutting tool differs from a lathe tool in Tool angles.
- Tool which is used in shaper to produce a small contoured surface is forming tool.
- In a shaper tool the amount of side clearance angle is only 2 to 3°.
- In a machining horizontal surface shaper operation, the amount of depth adjusted by micrometer dial.
- Maximum size of pillar drilling machine is about 50 mm.
- Diameter of T – bots of drilling machines usually ranges between 15 to 20 mm.
- Plain slot clamp of drilling machine is made with Mild steel.
- A pattern may be defined as a Replica or Facsimile model of the desired casting.
- Bench mounting is usually employed for small casting which are light in weight.
- Knurling is the process of embossing a diamond shaped pattern on the surface of a Work piece.
- Temperature of pouring metal is usually checked with Optical Pyrometer.
- The usual size range of Shaper in between170 to 900mm.
- For general purpose work, the included angle of the lathe centre is 60°.
- Diameter of T – bots of drilling machines usually ranges between15-20 mm.
- Pattern maker is basically a wood worker.
- The operation in which metal will remove in the form of chips by feeding the work against a grinding wheel is Grinding.
- The type of shaper mostly used in tool room work is Universal shaper.
- Universal chuck is also called as Self centring chuck.

- Half-nut mechanism makes the carriage to engage or disengage with the lead screw.
- Grooving is process of reducing the diameter of a work piece over a very narrow surface.
- The type of clamp used for quick adjustment of the work is U-Clamp.
- The type of holding device in drilling machine used for mass production is Drill jags.
- The Drilling machine suitable for drilling medium to large and heavy work pieces is radial drilling machine.
- The ferrous metals are those which have the iron as their main constituent, such as cast iron, wrought iron and steel.
- The non-ferrous metals are those which have a metal other than iron as their main constituent, such as copper, aluminium, brass, tin, zinc, etc.
- Strength is the ability of a material to resist the externally applied forces without breaking or yielding.
- The internal resistance offered by a part to an externally applied force is called stress.
- Stiffness is the ability of a material to resist deformation under stress.
- The modulus of elasticity is the measure of stiffness.
- Elasticity is the property of a material to regain its original shape after deformation when the external forces are removed.
- Elasticity is desirable for materials used in tools and machines.
- Steel is more elastic than rubber.
- Plasticity is property of a material which retains the deformation produced under load permanently.
- Ductility is the property of a material enabling it to be drawn into wire with the application of a tensile force.
- A ductile material must be both strong and plastic.
- The ductility is usually measured by the terms, percentage elongation and percentage reduction in area.
- Brittleness is the property of breaking of a material with little permanent distortion.

- Malleability is a special case of ductility which permits materials to be rolled or hammered into thin sheets.
- The impact strength of a material is a index of toughness.
- The property of a material which allows it to be drawn it into a smaller section is called ductility.
- The loss of strength in compression due to overloading is known as creep.
- The maximum strain energy that can be stored in a body is known as resilience.
- The total strain energy stored in a body s termed as resilience.
- Proof resilience per unit volume of a material is known as modulus of resilience.
- Proof stress is the stress causing a specific permanent deformation usually 0.1-0.2%.
- Thermal strain caused in the material of a composite body due to change in temp will be same magnitude.
- True stress is related to simple stress/ strain.
- Bulk modulus'k' in terms of modulus of elasticity (E) and poissons ratio (μ) is $E/3(1-2\mu)$.
- Value of factor of safety is greater than 1.
- Hooks law is truly valid upto proportional limit.
- In ductile material nominal breaking stress is lower than the true breaking stress.
- In ductile material ultimate stress is higher than the normal breaking stress but lower than true breaking stress.
- In a brass specimen subjected to tension, proof stress can be obtained in a stress strain diagram.
- Thermal change of length of metal is directly proportional to its thermal coefficient.
- Thermal strain of a body does not depend on length.

12

Instrumentation and Measurement System

Measurement also forms a basic element of any control process. The measurement of discrepancy between the actual and desired performance of a system is central to any control process. Electronic instrumentation provides the solution to many and varied problems of measurement and control. It is often required to change or transduce a measurand into a corresponding electrical signal. The elements which changes/converts the physical variable in electrical signal are termed as sensors or transducers. For a measurement to be useful it must be reliable. That is where question of accuracy or uncertainty of a measurement arises. This may be dealt with clear understanding of measurand and measurement system.

- The error which is repetitive in nature is systematic error.
- Zero error of micro meter is systematic error.
- Threshold of the instrument is defined as the smallest measurable input signal which can be detected by the instrument.
- The smallest change in the volume of input variable being measured that will cause a change in the output signal of the instrument is termed as resolution.
- Repeatability of the instrument with respect to a given fixed input is precision.
- The gradual departure of the instrument output caused by certain interfering input and component instabilities is termed as drift.
- Error caused by the art of measurement on the physical system being tested as reading errors.
- Maximum power is transmitted by an electrical transducer if the impedence of external load matches with internal impedence of the transducer.
- Static sensitivity of the instrument is the ratio of change in input to the corresponding change in the input variable.

- The desired input to the instrument may be constant or varying slowly with respect to time is static character.
- The desired input is not constant but varies rapidly with time is dynamic character.
- The quantity whose magnitude is unknown is known as measurand.
- The quantity whose magnitude is known is known as standard.
- The process of comparison of measurand with standard is known as measurement.
- The method in which unknown quantity is directly compared with standard is known as direct method of measurements.
- Primary measurements can be made by direct observation without involving any conversion of measured quantity.
- Secondary measurement involves one translation of measured quantity into change of length.
- Tertiary measurements involves two conversions of measured quantity.
- In deflection type measurements, deflection of the instrument provides a basis for determining the quantity under measurement.
- In null type instruments, zero or null indication leads to determination of magnitude of measured quantity.
- Information is the data or details relating to an object or event.
- Static error is the difference between measured value and true value of the quantity.
- Accuracy of an instrument is measured in terms of its error.
- Accuracy is the closeness with which an instrument reading approaches the true value of quantity being measured.
- Precision is the measure of reproducibility of measurements.
- Dead time is the time before the instrument begins to respond after the measured quantity has been changed.
- Semiconductor strain gauge is the most sensitive type of sensing element for strain measurement.
- LVDT used for displacement measurement is an externally power operated transducer.

- The most common transducer for shock and vibration measurement is piezo metric pick up.
- LVDT works on the principle of variable mutual induction.
- Electrical material has got the property of generating emf when subjected to mechanical strain.
- The calibration of strain gauge bridge circuit is carried out by applying a known mechanical strain on the active gauge.
- A solar cell is photo voltaic transducer.
- LVDT - linear variable differential transducer.
- Shaft encoder is the instrument for angular measurements.
- For dynamic pressures the piezo electric transducer is suitable .
- A thermostatic cutoff works on the principle of thermal expansion of metals.
- The basis for measuring the thermodynamic property, the temperature is given by Zeroth law of thermodynamics.
- According to Steffen Boltzmann law the amount of radiant energy per unit area is proportional to fourth power of absolute temp.
- Bimetallic strips made of two different materials bent during a rise in temp on the account of difference in coefficient of linear expansion.
- The principle of working the constant volume thermometer is based on boyle's law.
- The unit of Steffen Boltzmann constant is $W/cm\text{-}K^4$.
- Optical pyrometer is used to measure high temperature.
- At -40°c Fahrenheit and celcius scales coincide.
- Boiling point of water which is used as one of the fixed point in international practical temp in K is 372.45.
- The approximate range upto which ordinary mercury glass thermometer can be used is -20 to 340 °C.
- The primary transducer element in a pressure thermometer is bulb together with capillary tube.
- A thermocouple arrangement is to be used to measure a high temp of 1400°C point out the pair of thermocouple that would be most suitable for this application is platinum 13 % typr-R.

- The instrument which measures the temperature of source without direct contact is pyrometer.
- The sensing element of the industrial pressure thermometer is usually made of steel.
- Copper – corestantan thermocouple has the lowest temp measuring range.
- Resonant frequency of quartz crystal change with temp.
- Silicon based temp sensors has excellent linear characteristics.
- Turbine flow meter is suitable for flow totalization.
- An electromagnetic flow meter can be mounted in any direction.
- Electromagnetic flow meter is independent of fluid density.
- Rotameter is a variable area flow meter.
- Sliding vane flow mater is a positive displacement device.
- Turbine flow meter is used for clean fluids only.
- The tube of rotameter tapers down wards.
- Lobbed impeller flow meter is capable of giving the rate of flow as well as the totalization flow.
- A rotometer can be used in vertical orientation.
- The head loss of an orifice is greater than that of flow meter.
- Electromagnetic flow meters are used to measure the flow of conducting fluids in a plastic pipe.
- Hot wire anemometer is a device used to measure gas velocity.
- In rotameter the pressure drop remains nearly constant but area changes.
- The transducer preferred to measure highly fluctuating velocities is hot wire anemometer.
- In an electromagnetic flow meter the induced voltage is proportional to flow rate.
- Weir is used to measure the flow rate in an open channel.
- If a measurement tape is too long as compared to standard, the error will be known as instrumental error.
- The amount of moisture in air is measured by sling psychrometer.

- Thermistor can be used as thermal detector.
- Thermocouple can be used to give an indication for the temperature change.
- Thermocouples are generally used for temperature measurement upto 1600ºC.
- The function of reference electrode in a ph meter is to produce a constant voltage.
- A hydrometer can be used to measure the specific gravity of liquids.
- A pitot tube converts liquid head into pressure head.
- A LVDT has one primary coil and one secondary coil.
- A pirani guage is used to measure very low pressure.
- Precision of an instrument is in fact dependent on the repetability.
- One torr is a pressure equivalent to 1 mm of Hg.
- Bonded strain guage transducer is widely used for measurement of several physical variables like strains, torque, pressure, vibration etc.
- Selective radiation pyrometer is a temperature measuring device works on the principle of planks law.
- Electrical resistance of a thermometer decreases as temperature increases.
- The best use of iron constant thermocouple is in the temperature range of 63-1473K.
- The best use of resistance thermometer is in the range of below 630ºC.
- The least count of any instrument is taken as the resolution.
- The smallest measurable input above the zero value is called threshold.
- The largest change of the measurement to which the instrument does not respond is known as dead band.
- The maximum angle or distance through which any part of the mechanical system may be moved in one direction without causing motion of any part is backlash.
- A pirani gauge works on the principle of change of thermal conductivity of medium.
- Ionization gauge is an indirect method of pressure measurement.

- The error that always follows same definite mathematical or physical law is known as systematic errors.
- The elements that senses and converts the desired input into a more convenient and practicable form to be handled by the measurement system is transducer element.
- The element implied for giving the information of the measured variable in the quantitative form is data presentation element.
- The activity in which increasing the amplitude of the signal without affecting its wave form is called amplification.
- Instrumentation Some Relationships:

  Pg = Pa- Ps

  Pv = Ps- Pa

  Pg = Guage pressure, Pa = Absolute Pressure, Pv = Vaccum pressure

  Ps = Atmospheric pressure.
- Velocity pressure = total pressure - static pressure.
- The atmospheric pressure at sea level is 101.3 kN/m^2 or 760 mm of mercury.
- K = Boltzmann's constant = 1.38 X 10^{-23} J/°K.
- Bimetallic thermometers are used for measurement of temperatures between -40°C and 550°C.
- Invar is the most commonly used low expansion material for bimetallic thermometers.
- In bimetallic thermometers, the deflection of strip is directly proportional to temperature change and square of length of the strip.
- The sensitivity of microphone is a measure of voltage developed per unit sound pressure.

Important Scientic Instruments and their usage

Accumulator	: It is used to store electrical energy.
Altimeter	: It measures altitudes and is used in aircrafts.
Ammeter	: It measures strength of electric current (in amperes).
Anemometer	: It measures force and velocity of wind.

Audiometer	: It measures intensity of sound.
Audiphones	: It is used for improving imperfect sense of hearing.
Barograph	: It is used for continuous recording of atmospheric pressure.
Barometer	: It measures atmospheric pressure.
Binocular	: It is used to view distant objects.
Bolometer	: It measures heat radiation.
Calorimeter	: It measures quantity of heat.
Carburettor	: It is used in an internal combustion engine for charging air with petrol vapour.
Chronometer	: It determines the longitude of a place kept onboard ship.
Colorimeter	: An instrument for comparing intensities of colour.
Commutator	: An instrument to change or remove the direction of an electric current, in dynamo used to convert alternating current into direct current.
Cresco graph	: It measures the growth in plants.
Cyclotron	: A charged particle accelerator which can accelerate charged particles to high energies.
Dynamo	: It converts mechanical energy into electrical energy.
Dynamometer	: It measures force, torque and power.
Electroscope	: It detects presence of an electric charge.
Eudiometer	: A glass tube for measuring volume changes in chemical reactions between gases.
Fathometer	: It measures the depth of the ocean.
Galvanometer	: It measures the electric current of low magnitude.
Hydrometer	: It measures the specic gravity of liquids.
Hydrophone	: It measures sound under water.
Hygrometer	: It measures humidity in air.
Kymograph	: It graphically records physiological movements (Blood pressure and heart beat).
Lactometer	: It determines the purity of milk.

Manometer : It measures the pressure of gases.

Mariner's compass : It is an instrument used by the sailors to determine the direction.

Microphone : It converts the sound waves into electrical vibrations and to magnify the sound.

Microscope : It is used to obtain magnied view of small objects.

Odometer : An instrument by which the distance covered by wheeled vehicles is measured.

Phonograph : An instrument for producing sound.

Photometer : The instrument compares the luminous intensity of the source of light.

Potentiometer : It is used for comparing electromotive force of cells.

Pyrometer : It measures very high temperature.

Radar : Radio, angle, detection and range is used to detect the direction and range of an approaching aeroplane by means of radio micro waves.

Radiometer : It measures the emission of radiant energy.

Rain Gauge : An apparatus for recording rainfall at a particular place.

Rectier : An instrument used for the conversion of AC into DC.

Refractometer : It measures refractive index.

Saccharimeter : It measures the amount of sugar in the solution.

Salinometer : It determines salinity of solution.

Seismograph : It measures the intensity of earthquake shocks.

Spectrometer : It is an instrument for measuring the energy distribution of a particular type of radiation.

Spectroscope : An instrument used for spectrum analysis.

Speedometer : It is an instrument placed in a vehicle to record its speed.

Spherometer : It measures the curvatures of surfaces.

Sphygmomanometer : It measures blood pressure.

Stereoscope : It is used to view two dimensional pictures.

Stroboscope	: It is used to view rapidly moving objects.
Tachometer	: An instrument used in measuring speeds of aero planes and motor boats.
Telescope	: It views distant objects in space.
Theodolite	: It measures horizontal and vertical angles.
Thermometer	: This instrument is used for the measurement of temperatures.
Thermostat	: It regulates the temperature at a particular point.
Transistor	: A small device which may be used to amplify currents and perform other functions usually performed by a thermionic valve
Udometer	: It is used to measure the amount of liquid precipitation over a set period of time. It is also called Rain Gauge.
Vernier	: An adjustable scale for measuring small subdivisions of scale.
Viscometer	: It measures the viscosity of liquids.
Voltmeter	: It measures the electric potential difference between two points.

13

Testing and Evaluation in Farm Machinery and Power

- Spray Angle — The angle subtended at the final orifice by the edges of the spray pattern.
- Variation in discharge rate between first and last observation shall not be more than 5 percent.
- The volumetric efficiency requirement for roller vane type pump shall be minimum 80 percent.
- The volumetric efficiency of piston, plunger type pump shall be minimum 80 percent.
- If the droplet size is reduced from 2 mm to 1 mm, number of droplets produced will increase by 8 times from the same volume.

Target	Droplet sizes (Microns)
Flying Insects	10-50
Insects on foliage	30-50
Foliage	40-100
Soil application (avoidance to drift)	250-500

- The LAI ratio seldom exceeds about 6-7 depending upon the crop.

$$\text{Leaf Area Index (LAI)} = \frac{\text{Leaf Area}}{\text{Ground Area}}$$

- The height of cutter bar above ground level shall be at 150 mm in combine harvester.
- The maximum actuating force shall not be more than 400 N for hand operated and 600 N for foot operated parking brake device.
- Contact Angle - The forward angle between the horizontal ground and the line joining the shovel tip touching the ground and its centre when

shovel is fitted with tine and placed on its working position. (150-165 degree).

- Lift angle- The inclination of wing in the direction of travel when sweep is placed on flat surface and in working position.
- The shovel and sweep (carbon steel) shall have hardness in the range of 350 to 450 HB and Tine is made of mild steel.
- Spring index - Ratio of the mean diameter of the spring coil and the diameter of spring wire(4-5).
- Seeding Uniformity –

  a) Sticky Belt Method (10m long centre to centre)

  b) Sand Bed Method (5m long and 25cm depth).
- The size of the trial plot shall not be less than 0.1 ha and 0.2 ha for the animal drawn and tractor operated drills respectively.
- Drawbar Power- Power measured at the drawbar which can be sustained for at least 20s, or the time needed to cover a distance of at least 20m, whichever is longer.
- Measure inflation pressure with the tyre valve in the lowest position.
- The power rating of the tractor is usually stated as PTO. If the tractor is not fitted with a PTO capable of transmitting the full power of the engine, the power rating of the tractor shall be stated as the power measured at the drawbar.
- The minimum area of the plot for testing animal drawn disc harrow shall be 0.25 hectare and for tractor operated harrows, one hectare. The ratio of width and length of the plot should be, as far as possible, 1: 2.
- Soil pulverization - Measure the mean mass diameter (MMD) of soil aggregate which will be the index for soil pulverization.
- The slip of wheels or rubber tracks shall not exceed 15 percent and that of steel tracks shall not exceed 7 percent.
- The drawbar pull shall be 75 percent of the pull at maximum power in that gear.
- The minimum area of the plot for testing animal drawn disc harrow shall be 0.25 hectare and for tractor operated harrows, one hectare. The ratio of width and length of the plot should be, as far as possible, 1 : 2.
- Measure the mean mass diameter (MMD) of soil aggregate which will be the index for soil pulverization.

- The furrow openers of shovel and shoe type shall be hardened to have Brinell hardness between 350 and 450 HB.
- Transmission system of seed drill shall be sprocket and chain, or gear type.
- The variation in dropping of seed & fertilizer in different feeding outlets separately from average quantity obtained should not be more than 5 & 7.5% respectively.
- The percentage visible damage to seed after passing through metering device should not exceed 0.25%.
- The variation in seed dropping due to box filling at rated capacity shall not exceed 10%.
- Lug angle of power tiller cage wheels as per BIS specifications is 20°+1°.
- The variation in uniformity of seed dropping and placement at a forward speed of 1.5 kmph in animal drawn semi-automatic sugar cane planter should not be more than 2%.
- The cultivator shovel should have brinell hardness number ranging from 250-300.
- Recommended tractor speed of tilling with rotavator as per BIS standards is 2.5 kmph.
- The variation in dropping of seeds from each chute in semi-automatic potato planter from its average quantity obtained should not be more than 5%.
- The percentage of germ damage in potato planter should not exceed 1.5%.
- The variation in dropping of seed and fertilizer in different feeding outlets separately shall be not more than 7 and 12.5 percent respective y from the average quantity obtained.
- The variation in quantity dropped per hectare and quantity specified to be dropped at a particular setting shall be not more than 7 and 12.5 percent for seed and fertilizer respectively.
- The seed and fertilizer rate shall be easily adjustable up to 125 kg and 1000 kg per hectare respectively.
- The percentage of visible damage to seed in the drill shall not exceed 0.5 percent.
- The variation in dropping due to box filling at 1/4, 1/2 and 3/4 of rated capacity shall not exceed 10 percent.

- The variation in quantity of seed dropping due to change in speed shall not exceed 15 percent of the quantity dropped under the specified speed.
- The variation in quantity of seed per metre of row length shall not exceed by 10 percent.
- The drill shall be able to sow seed up to 100 mm deep and should be able to drop fertilizer at a minimum of 25 mm to the side of the seed.
- The draft of animal-&awn drill shall be not more than 125 kgf.
- The wheel slip at specified speed shall not exceed by 15 percent.
- The animal-drawn drill shall not have more than five furrow openers; and tractor-operated drill shall have 5 to 15 furrow openers.
- When the furrow openers are lowered to plain surface, the openers shall not deviate by more than 5 mm from the line of alignment vertically and horizontally.
- The weight of tractor-mounted drill including the weight of seed and fertilizer filled at rated capacity of box shall not exceed 300 N/kW drawbar power of the tractor recommended for the drill.
- Contact Angle: The forward angle between the horizontal ground and the line joining the shovel tip touching the ground and its centre when shovel is fitted with tine and placed on its working position.
- Lift Angle: The inclination of wing in the direction of travel when sweep is placed on flat surface and in working position.
- The shovel and sweep shall have hardness in the range of 350 to 450 HB.
- Contact Angle: the angle shall not differ by + or – 3 from the declared value. The angle may be between 150° to 165°.
- Cultivator – Mild steel .
- Seed Metering Mechanism for potato planter semi-automatic - Metering mechanism shall be of horizontally revolving ring type or of belt with cups.
- The variation in dropping of seed (**potato**) from each chute shall be not more than 5 percent from the average quantity obtained.
- The variation in quantity of seed dropped per hectare and quantity specified to be dropped at a particular setting shall be not more than 7 percent.
- The percentage of germ damage shall not exceed 0.5 percent.
- The variation in dropping due to box filling at 1/4, 1/2, and 3/4 of rated capacity shall not exceed by 10 percent.

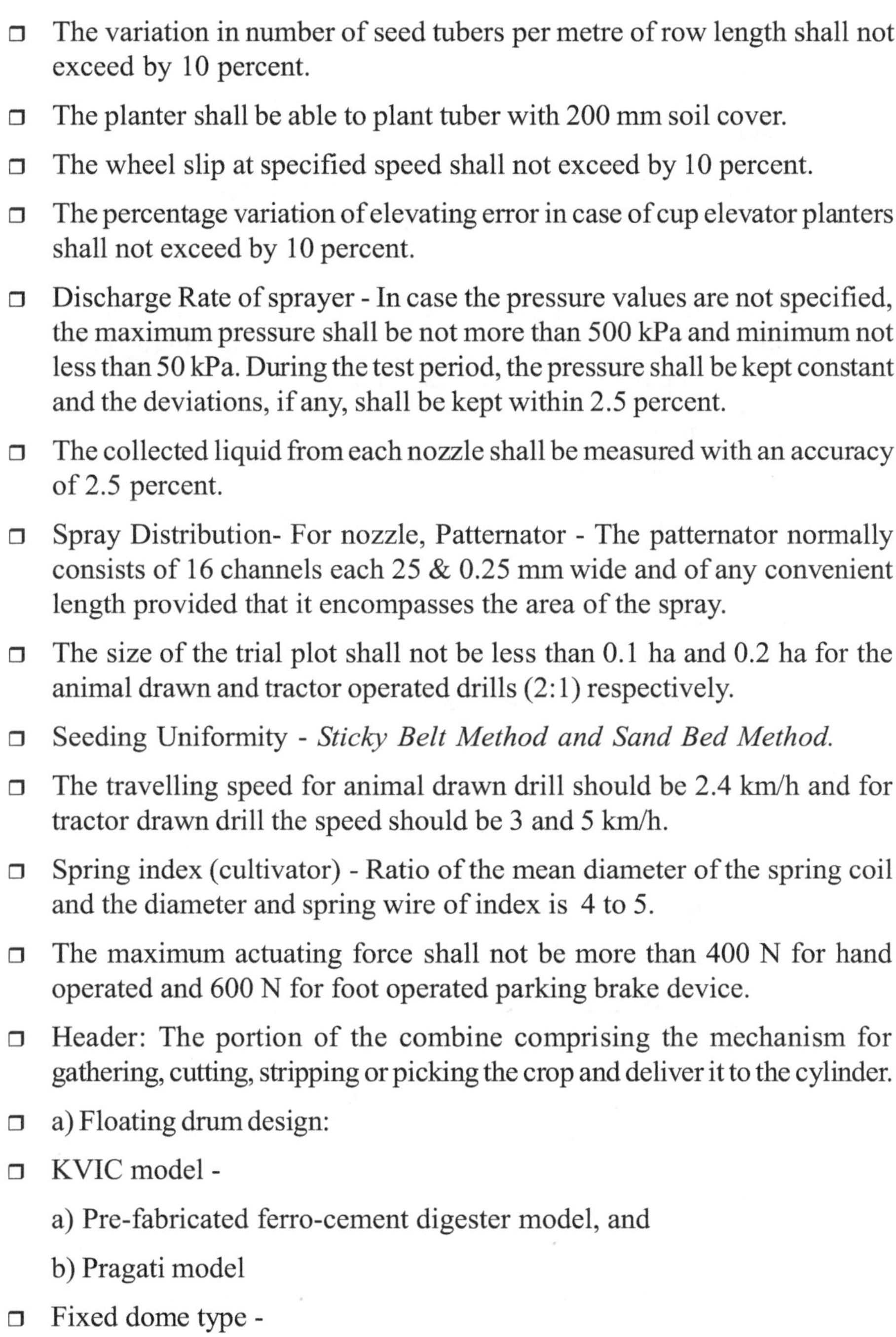

- The variation in number of seed tubers per metre of row length shall not exceed by 10 percent.
- The planter shall be able to plant tuber with 200 mm soil cover.
- The wheel slip at specified speed shall not exceed by 10 percent.
- The percentage variation of elevating error in case of cup elevator planters shall not exceed by 10 percent.
- Discharge Rate of sprayer - In case the pressure values are not specified, the maximum pressure shall be not more than 500 kPa and minimum not less than 50 kPa. During the test period, the pressure shall be kept constant and the deviations, if any, shall be kept within 2.5 percent.
- The collected liquid from each nozzle shall be measured with an accuracy of 2.5 percent.
- Spray Distribution- For nozzle, Patternator - The patternator normally consists of 16 channels each 25 & 0.25 mm wide and of any convenient length provided that it encompasses the area of the spray.
- The size of the trial plot shall not be less than 0.1 ha and 0.2 ha for the animal drawn and tractor operated drills (2:1) respectively.
- Seeding Uniformity - *Sticky Belt Method and Sand Bed Method.*
- The travelling speed for animal drawn drill should be 2.4 km/h and for tractor drawn drill the speed should be 3 and 5 km/h.
- Spring index (cultivator) - Ratio of the mean diameter of the spring coil and the diameter and spring wire of index is 4 to 5.
- The maximum actuating force shall not be more than 400 N for hand operated and 600 N for foot operated parking brake device.
- Header: The portion of the combine comprising the mechanism for gathering, cutting, stripping or picking the crop and deliver it to the cylinder.
- a) Floating drum design:
- KVIC model -

  a) Pre-fabricated ferro-cement digester model, and

  b) Pragati model
- Fixed dome type -

  a) Janta model, and

  b) Deenbandhu model

- It is a gaseous mixture containing about 60 percent methane, a high value fuel, about 10 percent carbon-dioxide and traces of other gases, such as ammonia and hydrogen sulphide, obtained from anaerobic fermentation of biomass.
- Anaerobic Fermentation: A biological conversion process carried out by Micro-organisms which convert bio-mass, in the absence of oxygen, to methane.
- The capacity of the gas plant is determine the estimated quantity of gas produced from the available quantity of bio-mass and consumption of gas for different purposes based on the following data:
- Gas consumption:

  For stove - 0.45 m^3 per hour

  For lamp - 0.13 m^3 per mantle hour (100 candle power).
- For dual fuel engine (biogas /diesel): 0.45 m^3 per hp per hour (based on 80 percent replacement of diesel oil).
- As per BIS standard, the power test for tractor pto includes Maximum power, varying load and varying speed test.
- Drawbar Power : Power measured at the drawbar which can be sustained for at least 20 sec, or the time needed to cover a distance of at least 20m, whichever is longer.
- The fuel consumption shall be measured when the tractor traverses a straight track for a minimum distance of 100 m.
- For wheeled tractors, the following formula

  $H^{max} = 0\ 8 \times W \times Z / F$

  where,

  W is the static load exerted by the front

  wheels on the ground, In newtons

  Z is the wheelbase, In milhmetres,

  F is the drawbar pull, In newtons, and

  H is the static height of the hne of pull above the ground, In milhmetres.
- The minimum area of the plot for testing animal drawn disc harrow shall be 0.25 hectare and for tractor operated harrows, one hectare. The ratio of width and length of the plot should be, as far as possible, 1 : 2.

- Baffle-Plate: An element placed near the rear beater to prevent grain from being thrown to the straw walkers.
- Blower: A rotary device which produces a draught or air across the chaffer and cleaning sieve(s) to blow away the material lighter than grains.
- Lifting Speed Index: The ratio between the peripheral speed of lifting chains and the operating speed of combines.
- Processing loss The loss of grains in terms of damaged, unthreshed and threshed obtained after completion of threshing, separating and cleaning operations.
- a) Cylinder loss - The loss of unthreshed heads and damaged grains passing out of threshing cylinder, rock and shoe loss.

  b) Rack loss - The threshed grains passing out in the straw.

  c) Shoe loss - The threshed grains blown or carried out with the chaff
- Size of Plot (Paddy weeder) - For uniformity of working, the minimum size of fieldnmay be fixed as 10 x 10 metres.
- The depth of water to be allowed at the time of weeding may be 2.5 mm for facilitating the cleaning operation and effective elimination of weed.
- The size of the drill shall be expressed by the number of seed furrow openers and the maximum spacing in milli metres between two adjacent furrow openers.

  Puddling Index - It is the ratio between volume of settled soil and total volume of sample and it is given by the formula:

  $$\text{Puddling Index} = \frac{Vs}{V} \times 100$$

  where,

  Puddling Index = v X 100

  Vs = volume of settled soil, and

  V = total volume of sample.
- The minimum area of the plot for testing of animal drawn puddler shall be 0.25 hectare and for tractor-operated puddlers, one hectare. The ratio of width and length of the plot shall be as far as possible 1 : 2.
- Minimum 5 to 10 cm of water column shall be maintained in the test plot.

TRACTOR

1.	Test Code for Agricultural Tractors	IS :	5994-1995
2.	PTO performance	IS :	12036-1995
3.	Belt-pulley performance	IS :	12036-1987
4.	Drawbar performance	IS :	12226-1987
5.	Hydraulic power and lifting capacity	IS :	12224-1987
6.	Turning Ability	IS :	11859-1986
7.	Position of centre of gravity	IS :	10743-1983
8.	Operators field of vision	IS :	11442-1985
9.	Brake Test	IS :	12061-1987
10.	Test of air cleaner	IS :	5995-1995
11.	Smoke Level	IS :	12062 1987
12.	Noise measurement	IS :	12180-1987
13.	Vibration measurement	IS :	5994-1995
14.	Haulage and field test	IS :	9253-1987
15.	Component/assembly inspection	IS :	5994-1995
16.	Selected performance characteristics of Agri. Tractors	IS :	12207-1987
	Technical requirements for operators seat of Agri.Tractors	IS :	12343-1988
	Safety and comfort	IS :	12239-1995
		•	(Part I& II)
	Symbols for operators controls on Agri. Tractors and Farm Machinery	IS :	6283-1971

COMBINE HARVESTER

1.	Test Code for Combine Harvester- Thresher	IS :	8122-1981&94
		•	(Part I & II)
2.	Methods of test for Internal Combustion Engines	IS :	10000-1991
3.	Specification for knife section for harvesting machines	IS :	6025-198

POWER TILLER

1.	Test code for Power tiller	IS	:	9935-1988
2.	Field performance and Haulage test	IS	:	9980-1988
3.	Selected performance characteristics	IS	:	13539-1992

14
Formulae

Area of ploughing

A= W x Dc

A – area of ploughing, m^2

W – width of cut, m

Dc – depth of cut, m

Draft of Plow

D = Ac δs and WsQ

D – draft of plow, kg

Ac – area of cut, m^2

δs – specific resistance of soil, kg/m^2

Drawbar Horsepower

$$DBHP = \frac{D \times S}{76.2}$$

DHP – drawbar horsepower

D – draft of implement, kg

S – speed of implement, m/s

Theoretical Field Capacity

$$TFC = \frac{W \times S}{10}$$

(ha/h)

TFC = Theoretical Field Capacity

W = Rated Width of implement, m

S = speed of implement, km/h

C = effective field capacity, ha/h

Effective Field Capacity

$$C = \frac{SW}{10} \times \frac{Ef}{100}$$

W = Rated Width of implement, m

S = Speed of implement, km/h

E_f = field efficiency, %

Field Efficiency

$$E_f = \frac{C_e}{C_t} \times 100$$

E_f – field efficiency, %

Ce – effective field capacity, ha/h

Ct – theoretical field capacity, ha/h

Number of Implement Unit

$$NI = \frac{Af}{To\,Ce}$$

NI – number of implement units

Af– area of the farm, hectares

To – total operating time to finish operation,hours

Ce – effective field capacity of implement,ha/hr

Width of Cut of Disc Plow

$$W = \frac{0.95\ N\ S + D}{1000}$$

W - width of cut, m

N - number of disk

S - disk spacing, mm

D - diameter of disk, mm

Width of Cut of Disc Harrow (Single Action)

$$W = \frac{0.95\ N\ S + 0.3\ D}{1000}$$

W - width of cut, m

N - number of disk

S - disk spacing, mm

D - diameter of disk, mm

Width of Cut of Disc Harrow (Tandem Type) (Double Action)

$W = 0.95\ N\ S + 1.2\ D/1000$

W - width of cut, m

N - number of disk

S - disk spacing, mm

D - diameter of disk, mm

Draft of Moldboard Plow

$D = 7.0 + 0.049\ S^2$: silty clay

$D = 6.0 + 0.053\ S^2$: clay loam

$D = 3.0 + 0.021\ S^2$: loam

$D = 3.0 + 0.056\ S^2$: sandy silt

$D = 2.8 + 0.013\ S^2$: sandy loam

$D = 2.0 + 0.013\ S^2$: sand

D - unit draft of implement, N/cm2

S - implement speed, kph

Tractor

Wheel Power

$Pw = \eta Pe$

Pw - wheel power, kw

Pe - engine power, kw

η-mechanical efficiency, 0.75 to 0.95

PTO Power

PPTO= ηPe

PPTO – PTO horsepower, kw

Pe – engine power, kw

η-mechanical efficiency, 0.75 to 0.95

Wheel Axle Torque

T = 1000 N/2 πn

T – wheel axle torque, N-m

N – wheel axle power, kw

n – speed of the wheel axle, rpm

Traction Efficiency

çd = Pd/ Pw

ηd – traction efficiency, %

Pd – drawbar power, kw

Pw – wheel power, kw

Running Resistance

R = Cr W

R – rolling resistance, kgf

Cr – coefficient of rolling resistance0.01 to 0.4 forwheel type and 0.05 to 0.12 for track type

W – trator weight, kg

Drive Wheel or Track Slippage

% Slip = 100x (R – r)/ r

% Slip – percent wheel slip,

% R – total drive wheel revolution count to traversethe drawbar runway under no load, rev

r – total drive wheel revolution count to traversethe drawbar runway under load, rev

Travel Reduction or Slip

$$S = 100\frac{An - Al}{Al}$$

Al S – slip,

% An – tract revolution under no load condition, m

Al – tract revolution under load condition, m

Stability Factor

$$K = \frac{FwWb}{Ph}$$

K – stability factor, 1.25 min

Fw – static front end weight, kg

Wb – wheel base,

P – maximum drawbar pull parallel to ground, kg

h – height of static line of pull perpendicular to ground

Drawbar Power

$DP_{kw} = (D\ S) / 3.6$

DHP - drawbar power, kW

D - force measured, kN

S - forward speed, km/hr

Thermal efficiency $= \frac{\text{available useful work}}{\text{heating value of supplied fuel}} \times 100$

Indicated thermal efficiency $= \frac{\text{ip(kj/s)}}{\text{mass of fuel per second} \times \text{calorific value of fuel}}$

Brake thermal efficiency $= \frac{\text{bp}}{\text{mass of fuel per second} \times \text{calorific value of fuel}}$

Firing interval $= \frac{\text{BHP} \times 75 \times 60}{L \times A \times N \times \frac{n}{2}}$ For four stroke engine

Indicated horse power $IP = \frac{PLAN}{60{,}000} \times \frac{n}{2}[SI]$

$IHP = \frac{PLAN}{4500} \times \frac{n}{2} Mks$

Indicated mean effective pressure $P_{im} = \frac{\text{Area of indicated diagram}}{\text{Length of the indicated}}$

Brake horse power

$B.H.P = \frac{I.P.H \times Mechanical\ of\ ficiency}{100}$

$B.H.P = \frac{I.H.P}{Thermal\ of\ ficiency\ of\ engine}$

Drawbar horse power $Dbhp = \frac{\text{B.H.P}}{\text{Thermal of ficiency}}$

Mechanical efficiency $= \frac{\text{B.H.P}}{\text{I.H.P}} \times 100$

Compression ratio $= \frac{\text{Total cylinder volume}}{\text{clear ance volume}}$

Piston speed $= \frac{2 \times L \times N}{100}$ merter/minute

Displacement volume

$= A \times L \times N \times \frac{n}{2}$ For four stroke engine

$= A \times L \times N \times n$ For two stroke engine

Brake mean effective pressure $= \frac{\text{BHP} \times 75 \times 60}{L \times A \times N \times \frac{n}{2}}$ For four stroke engine

Brake thermal efficiency $= \frac{641 \times 10^2}{\text{S.F.C}\left(\frac{\text{kg}}{\text{bhp hr}}\right) \times \text{calorific value}\left[\frac{\text{Kcal}}{\text{hr}}\right]^{\text{unit}}}$

API degree of oil

Relation between API gravity scale and specific gravity scale is

$$\text{API degrees} = \left[\frac{141.5}{\text{Specific gravity at } 15.6\,^{\circ}\text{C}}\right] - 131.5$$

Forces resulting from piston assembly movement

$$F = M\theta^2 r\left(\cos\theta + \frac{r}{1}\cos 2\theta\right)$$

M = reciprocating mass

ω = angular velocity of crankshaft

r = radius

l = connecting rod length

= angle between connecting rod axis and cylinder axis

Air flow requirement of four stroke cycle engine

$$Q(m^3/s) = \frac{(\text{engine displcament (liters)} \times \text{rpm} \times \text{volumetric efficiency})}{1.2\times10^5}$$

Heating value of fuel

$$\text{Heat value} = \text{Quantity of fuel} \times \text{Calorific value fuel}$$

Absolute viscosity

$$\mu = \frac{\pi p r^4 t}{8\text{vl}}$$

μ- Absolute viscosity, poises

p- pressure difference in dynes/cm^2

r- radius of tube,cmt- time in seconds

v- volume of liquid,cm^3

l- length of tube,cm

Kinematic viscosity

$$\text{Kinematic viscosity} = \frac{\text{Absolute viscocity}}{\text{mass density}}$$

Viscosity index

$$= \frac{\text{L-U}}{\text{L-H}} \times 100$$

L - viscosity at 30°C of L and that has an arbitrary index of zero

H – Viscosity of H which have VI scale of 100

U - Viscosity of 38 °C of the where index is unkwon

Relative efficiency

$$\frac{\text{Indicated thermal efficiency}}{\text{Air standard effiency}}$$

Governor regulation

$R = \frac{\Delta N}{} \times 100$

R- Regulation,%

N- Difference in speed

$\bar{N}$- Average speed

Torque developed by disc clutch or brake

$T = Ffr_m N$

F= effective clamping force

f = coefficient of friction

r_m= mean radius of the clutch facings

N= number of torque transmitting surfaces

Heat generated by a clutch

$$Q = \frac{TNt}{9.5493}$$

Q= heat generated in slipping clutch, J

T= the average slip torque, N-m

N= the average slip, rpm

t= the total slip time, seconds

Weight transfer

$$= \frac{\text{Pull kg} \times \text{Hitch height in meter}}{\text{wheel base in meter}}$$

Wheel slip

$$S = 1 - \frac{V_a}{V_t}$$

S= wheel slip or travel reduction

V_a= actual travel speed

V_t = theoretical wheel speed

Rim pull (kg)

$$= \frac{170 \times HP \times efficiency}{speed(km/hr) \times 100}$$

Rolling resistance

$$= \frac{2}{(n+1)(K_c + BK_{Æ})1/n} \times \left(\frac{W}{2L}\right)\frac{n+1}{n}$$

n- coefficient of wheel shrinkage,

B - width of each track

W - vertical load over shear area

L - length of track in contact

Frequency

$$F = \frac{1}{T}$$

T- time

Kinetic energy for a wheeled tractor

$$KE = \frac{1.1}{2} Mv^2$$

M = mass of the tractor

v = velocity

Sound pressure level(SPL)

$$SPL = 20 \log \frac{P}{P_O} dB$$

SPL= sound pressure level, decibels

P= measured RMS sound pressure, N/ m^2

P_0= reference sound pressure, N/m^2

Log = logarithamic to the base 10

Transmissibility

$$\text{Transmissibility} = \frac{\text{output vibration intensity}}{\text{input vibration intensity}}$$

Damping ratio

$$\xi = \frac{C}{C_c}$$

ξ= damping ratio

C= seat suspension damping ratio, N/ m.s

C_c= critical damping rate, N/m.s

C_c= 2Mω_sM= mass of seat and operator, kg

Undamped natural frequency of the operator seat

$$\omega_S = \sqrt{\frac{k}{M}}$$

K = spring rate of seating system, N/m

ω_s= undamped natural frequency of the operator seat, rad/s

Critical damping rate

$$\varepsilon = \frac{C}{C_c}$$

C - seat suspension damping rate, N/m.s

C_c – Critical damping ratem, N/m.s

Torque

$T = \mu . F . r_m . n$

μ–coefficient of friction

F- effective clamping force

r_m – Mean radius of the clutch facing

n- No. of torque transmitting surfaces

Drawbar power	$=\frac{D.S}{3.6}$ D- Draft, KN S- Forward speed, kmph
Traction efficiency	$TE=\frac{\text{Drawbar pull}}{\text{Weight of tractor}}\times 100$
Average normal pressure for pneumatic tyre tractor	$P=\frac{W}{0.78bl}$ W- tractor weight kg b- width of the tyre in contact with soil l- length of tractor tyre in contact with soil
Gross traction force	$F = AC + W\tan\varnothing$ $\varnothing$-Angle of internal friction A-Area of contact C- Cohesion of soil
Pull (horizontal force)	$TF = W\left(\frac{1.2}{C_n}+0.04\right)$ C_n–Wheel numeric$=\frac{\text{Cl bd}}{W}$ CI- Cone index measured with a cone penetrometer TF- Towed force
Effective field capacity (ha/h)	$=\frac{SW}{1000}\times E_f$ S-speed of travel kmph W- width of implement in meter Ef- field efficiency in per cent
Draft	D= P cosθ D- draft in kgf P- Pull in kg θ- angle b/w line of pull and direction of motion
Unit draft	=Pcosθsinθ
Side draft	$=\frac{\text{Draft}}{\text{cross section area of furrow}}$

Relation between draft and speed

D_s = Draft at speed S

D_r = Draft at reference speed (4.83km/h)

S = speed, in kilometres per hour

Power

= Draft(N)Speed (m/s)

$$HP=\frac{\text{draft kg}\times\text{speed m/s}}{75}$$

Internal friction

$S= C+N\tan\Phi$

S- shear stress at soil failure surface,kg/cm^2

C- cohesion of soil

N- normal stress on the plane of shear failure

Φ- angle of internal friction

Friction (μ)

$$\frac{F}{N}=\tan\theta$$

μ– co effiecient of friction

F- Frictional force

N- Normal force perpendicular to surface

θ – angle of internal friction

Speed and draft

$D_s=D_o+KS^2$

D_s–Draft at speed S

D_o – Static component of draft

S- forward speed

K- Constant, depending on types of implement and soil conditions

Shearing stress at soil failure

$\tau=C+\sigma\tan\varphi$

τ= shearing stress at soil failure

C = cohesion

σ = stress normal to the plane of shear failure

Φ= angle of internal friction

Coefficient of friction

$$\mu=\frac{F}{N}=\tan\psi$$

μ - Coefficient of friction

F-frictional force tangent to the soil

surface

N - normal force

ψ - friction angle

Velocity ratio

$$VR = \frac{N_1}{N_2} = \frac{D_2}{D_1}$$

N_1 –No. of rev of driving pulley

N_2 – No. of rev of driven pulley

D_1 – Dia of driving pulley

D_2 – Dia of driven pulley

Seed rate (kg/ha)

$$\frac{\text{Seed collected kg}}{\text{Area ha}}$$

Spraying and dusting equipment

Field capacity

$$\text{Area covered (ha/h)} = \frac{N \times S \times W}{1000} \times$$

N- No. of nozzles

W- spacing between nozzles,m

S- Speed of travel, kmph

- spraying efficiency %

Application rate Spray volume

$$Q = C_d \times a\sqrt{2gh}$$

$\sqrt{2gh}$ -pressurwe developing in the nozzle,m

a - Projected area of the nozzle, m^2

C_d – Discharge coefficient

Time taken for spraying

$$\frac{\text{Total discharge volume}}{\text{Discharge rate}}$$

Discharge

Q= Area covered ×Application rate

Calculation for threshing equipment

1. Total grain input (A)

A=B+C+D

B-Quantity of grains collected at main outlets per unit time

C- Quantity of clean grains from sieve over flow, sieve underflow and at outlet per unit time

D- Unthreshed grains collected from all outlets per unit time

2. Percentage of broken grains

$$=\frac{E}{F}\times 100$$

E- quantity of broken grains in the sample at grain outlet

F- Total quantity of sample taken at main grain outlet

3. Percentage of blown grains

$$=\frac{G}{A}\times 100$$

G- quantity of clean grains obtained at straw outlet per unit time

A- Total grain input

4. Percentage of unhreshed grain

$$=\frac{H}{A}\times 100$$

H- quantity of unthreshed grains obtained from all outlets per unit time

5. Cleaning efficiency

$$c=\frac{M}{F}$$

M- quantity of clean grains obtained from the sample taken at main grain outlet

F- Total quantity of sample taken at main grain outlet

Chaff cutter Capacity of chaff cutter

$$T_t = 6\times 10^{-9}\ DALNR$$

Or

$$C = WHLNRK$$

T_t= theoretical capacity of chopper, tons per hour

D= density of forage as it passes between the feed rolls, in kilogram per cubic meter

A = throat area, cm^2

L = Theoretical length of cut, mm

N= number of cutters on cutter head

R = Speed of cutter head

Length of chaff

$$=\frac{\pi D}{n}\times\frac{N'}{N}$$

D- Dia of feed roller cmn- No. of knives

N'- Rev. of feed roller

N- Rev. of flywheel

Belt drive Relation between tensions

$\frac{T_1}{T_2} = e^{\mu\theta}$

T_1 – tensions on slack side

T_2 – Tensions on tight side

μ- coefficient of friction

θ– angle of lap of belt over pulley

Tension of belt due to centrifugal force

$T_c = mV^2$

T_c = Centrifugal tension, N

V= belt mass per meter of belt length

Average service life of belt

$$\text{Life (in hours)} = \frac{\text{(Belt length, in millimeters)} \times 100}{\text{total fatigue rate}}$$

Length of open belt

$$L = \frac{\pi}{2}(D_1 + D_2) + \frac{(D_1 + D_2)^2}{4G} + 2C$$

Length of cross belt

$$L = \frac{\pi}{2}(D_1 + D_2) + \frac{(D_1 + D_2)^2}{4G} + 2C$$

L=length of belt(open/cross)

D_1 – Dia of driver pulley,cm

D_2 – Dia of dribvem pulley

C- Distance between centre of both pulley, m

D-

Length of V belt

$$L = 2C + 1.57(D + d) + \frac{(D + d)^2}{4C}$$

$$C + \frac{1}{2}(D + d) + 3t$$

L= length of belt,m

E- Dia of driver pulley

d- dia of driven pulley

C- distance b/w centre of both pulley

t- belt thickness mm

Belt life in years

$$\frac{\text{Belt length in mm}}{\text{total fatique rate}} \times 100$$

Soil inversion $\frac{\text{No.of weeds before-No.of weeds after ploughing}}{\text{No.of weeds before ploughing}}$

Puddling index $\frac{V_S}{V}\times 100$

Vs- Volume of settled soil

V- Total Volume of soil

Diameter of disc =3.5×depth of ploughing

Utility index $K=\frac{\text{Number of working hours}}{\text{years}}\times 1000$

Heat balance equation S (storage) = M (metabolism) –E (evaporation) ±R (radiation) ±C (convection) – W (work accomplished)

Power tiller

Field Efficiency

Feff $Feff=\frac{EFC}{TFC}\times 100$

Feff–field efficiency, %

EFC–effective field capacity, ha/hr

TFC – theoretical field capacity, ha/hr

Fuel Consumption

$FC=\frac{v}{t}$

FC – fuel consumption, lph

V – volume of fuel consumed, L

t – total operating time, h

Axle/Rotary Shaft Torque

T = F L

T – shaft torque, kg-m

F – axle or rotary shaft load, kg

L – length of pony brake arm, m

Axle/Rotary Shaft Power

P = Ft N / 1340

P – shaft power, KW

Ft – total axle or rotary shaft load, kg

N – speed of axle or rotary shaft, rpm

Specified Fuel Consumption

SFC = Fc Pf / P

SFC – specific fuel consumption, (g/KW-h)

Fc – fuel consumption, L/h

Pf – density of fuel, g/h

P – axle or rotary shaft power, KW

Sowing

Nominal Working Width

$W = n\ d$

W - working width, m
n - number of rows
d - row spacing, m

Effective Diameter of GroundWheel

$De = d / \pi N$

De - effective diameter of ground wheel under load, m
d - distance for a given N, m
N - number of revolution, rpm

Delivery Rate

$Q = L\ 10{,}000 / \pi De\ N\ W$

Q - delivery rate, kg/ha
L - delivery for a given N, kg
De - effective diameter of ground wheel under load, m
N – number of revolution, rpm
W - working with, m

Delivery Rate (PTO-DrivenMachine)

$Q = L\ 10{,}000 / v\ t\ W$

Q - delivery rate, kg/ha
L - delivery for a given N, kg
v - tractor speed, m/s
t – time for measuring delivery, s
W - working with, m

Effective Field Capacity

$efc = A / t$

efc- effective field capacity, m2/h
A - area covered, m2
t – time used during operation, hr

Theoretical Field Capacity

$tfc = 0.36\ w\ v$

tfc- theoretical field capacity, m2/hr
w - working width, m
v - speed of operation, m/s

Field Efficiency

$Fe = (efc / tfc)\ 100$

Fe - field efficiency, %

efc- effective field capacity, m2/hr
tfc– theoretical field capacity, m2/hr

No. of Hills Planted

Hn = A 10,000 / SrSh

Hn- number of hills
A - area planted, hectares
Sr- row spacing, m
Sh- hill spacing, m

Wheel Slip

$$Ws = \frac{No - N1}{No} \times 100$$

Ws- wheel slip, %

No - sum of the revolutions of the driving wheelwithout load, rev

Nl- sum of the revolutions of all driving wheel withload, rev

Distance per Hill

Dph= SrπDg / Nc

Dph- distance per hill, mm
Sr- speed ratio of ground wheel and seed plate
Dg - diameter of the ground wheel, mm
Nc- number of cells in the seed plate

Speed Ratio of Ground Wheel andMetering Device

R = NcHs / Cgw

R - speed ratio
Nc- number of cells
Hs - hill spacing, m
Cgw- circumference of ground wheel, m

Total Weight of Seeds

TWs = NhNshSw / 1000 E

TWs - total weight of seeds needed, kg
Nh- number of hills
Nsh– number of seeds per hill
Sw- specific weight of seeds, g/seeds
E - emergence, decimal

Gear Ratio

GR = Tn/ Tr

GR - gear ratio
Tn- number of teeth of driven gear
Tr- number of teeth of driver gear

Design Power (Helical and SpurGears)

Pd = Pt (SFlo + SFlu)

Pd - design power, kW
Pt - power to be transmitted, kw
SFlo-service factor for the type of load, 1.0-1.8
SFlu-service factor for type of lubrication, 0.1-0.7

Center Distance

CD = M (t1 + t2) / 2

CD - center distance
M - module
t1 - number of teeth of the driven gear
t2 - number of teeth of the driver gear

Design Power (Straight BevelGear)

Pd = Pt SF / LDF

Pd- design power, KW
Pt - power to be transmitted, KW
SF – service factor, 1 to 2.5
LDF – load distribution factor, 1.0 to 1.4

Driver Gear Pitch Angle (StraightBevel Gear)

$\gamma = \tan^{-1} t1 / t2$

γ - pitch angle for the driver gear, deg
t1 – number of teeth of the driver gear
t2 – number of teeth of the driven gear

Driven Gear Pitch Angle (StraightBevel)

$\Gamma = 90° - \gamma$

Γ - pitch angle for the driven gear, deg
γ - pitch angle for the driver gear, deg

Wind Energy

Wind Power

Pw= ½ ρArV3

Pw – wind power, watts
ρ - air density, 1.25 kg/m3
Ar – rotor area, m2
V – velocity of the wind, m/s

Performance Coefficient

Pshaft = Cp½ ρA V3

Pshaft – power at the rotor shaft, watts
Cp– power coefficient, 0.17 to 0.47
ρ - air density, 1.25 kg/m3
A – rotor area, m2
V – wind velocity, m/s

Tip-Speed Ratio

$\lambda = 2\ \pi R\ N / V$

λ - tips-speed ratio, decimal
R – rotor radius, m
N – rotor speed, rps
V – wind velocity, m/s

Hydraulic Power

Ph= ρw g Q H

Ph– hydraulic power, watts
ρw – water density, 1000 kg/m3
g – gravitational acceleration, 9.8 m/s
Q – water flow rate, m3/s
H – lifting head, m

Windpump Rotor Diameter

Dr = (8 Ph/ πρwξV3)1/2

Dr – rotor diameter, m
Ph– hydraulic power, watts
ρw – density of water, 1000 kg/m3
ξ - overall system efficiency, 0.1
V – wind velocity, m/s

Conversions of units

Length	Area
1 ft = 12 inches	1 acre = 0.4047 hectare
1 yard = 3 feet	1 ha = 2.47 acre
1 mi = 5280 feet	1 ft2 = 144 in.2
1 cm = 0.3937 inch	1 acre = 43,560 ft2
1 inch = 2.54 cm	1 mi2 = 650 acres
1 m = 3.28 feet	1 m2 = 10.76 ft2
1 cm = 104 microns	1 ft2 = 929 cm2
1 mi = 1.609 km	1 in.2 = 6.452 cm2

Volume
1 liter = 1000 cc
= 0.2642 gal
= 61.025 in.3
= 103 cm3
1 ft3 = 144 in.3
= 7.482 gal
= 28.317 liter
= 28,317 cm3
1 acre-ft = 43,560 ft3
1 gal = 3.7854 liter
= 231 in3
= 8 pint
1 m3 = 35.31 ft3
= 103 liter

Density
1 lb/in.3 = 1728 lb/ft3
1 slug/ft3 = 32.174 lb/ft3
= 0.51538 gm/cm3
1 lb/ ft3 = 16.018 kg/m3
1 gm/cm3 = 1000 kg/m3

Angular 2π = 6.2832 radian
1 rad = 57.3 deg
1 rev = 2 π
1 rpm = 2 πrad/min
1 rad/sec = 9.549 rpm

Time 1 min = 60 seconds
1 hour = 3600 seconds
= 60 min
1 day = 24 hours

Speed
1 mph = 88 fpm
= 0.44704 m/s
= 1.467 fps
1 fps = 0.6818 mph
= 0.3048 m/s
1 knot = 0.5144 m/s
= 1.152 mph
1 m/s = 3.6 kph
= 2.24 mph
= 3.28 fps

Force, Mass
1 lb = 16 oz
= 444,820 dynes
= 32.174 poundals
= 4.4482 N
= 7000 grains
= 453.6 g
1 slug = 32.174 lb
= 14.594 kg
= 14.594 kg
1 kg = 2.205 lb
= 9.80665 N
= 1 kilopond

1 kip = 1000 lb
1 g = 980.665 dynes
1 ton = 2000 lb
= 907.18 kg
1 oz = 28.35 gm
1 metric ton = 1000 kg
1 Newton = 9.8 kgf
= 0.225 lbf

Pressure
1 atm = 1.033 bar
= 33.90 ft of water (at 4 °C)
= 10.33 m of water (at 4°C)
= 14.7 psi
= 101,325 N/m2
= 29.921 in. Hg (0°C)
= 33.934 ft H2O (60°F)
= 760 mm Hg (O°C)
= 406.79 in. H2O (39.2°F)
= 1.0332 kg/cm2
1 bar = 10 m of water
1 mm Hg = 13.6 kg (0°C)
1 psi = 27.684 inches of water
= 2.036 inches mercury
= 51.715 mm Hg (0 C)
= 0.0731 kg/cm2
1 psf = 47.88 N/m3
1 in. Hg = 13.57 in. H2O (60°F) (60°F)
= 0.4898 psi
1 N/m2 = 0.1 dyne/cm2
1 in H20 = 0.0361 psi
= 0.0736 inches mercury

Energy
1 Btu = 778.16 ft-lb
= 251.98 cal
= 1.055 kJ
1 hp-hr = 2544.4 Btu
1 J = 1 wt-s
= 1 N-m
= 0.01 bar-dm3
1 hp-s = 550 ft-lb
1 hp-min = 42.4 Btu
= 33,000 ft-lb1 kw-hr
= 3412.2 Btu
= 3600 kJ
1 kJ = 1 kw-s
= 101.92 kg-m
kcal/gmole = 1800 Btu/pmole
1 wt-s = 1 V-amp
1 kw-s = 737.562 ft-lb
1 kw-min = 56.87 Btu
1 atm-ft3 = 2.7194 Btu
1 J = 107 ergs1 ft-lb
= 1.3558 J1 kcal
= 4.1668 kJ1 hp
= 0.746 kw1 kW
= 1.34 hp
= 1.32 cv metric horsepower in French
1 PS = 0.986 Hp
1 wt-hr = 860 cal

Entropy, Specific Heat,
Gas Constant
1 cal/g-°K = 1 Btu/lb-°R
1 kcal/kg-°K = 1 kcal/kg-°R
1 Btu/lb-°R = 4.187 kJ/kg-°K

Standard Gravity g,
(as conversion unit)
1 slug = 32.174 fps2-lb

1 psin = 388.1 ips2-lb
1 s2-kg = 9.80665 N-m
1 s2-gm = 980.665 cm-dynes

Universal Gas Constant
1 pmole-°R = 1545.32 ft-lb
= 0.7302 atm-ft3
= 1.9859 Btu
= 10.731 psi-ft31 kgmole-°K
= 8.3143 kJ
= 0.08206 atm-m31 gmole-°K
= 82.057 atm-cm3
= 1.9859 cal
= 83.143 bar-cm3
= 8.3143 J
= 8.3149 x 107 erg
= 0.083143 bar-liter

Units of pressure
1 inch of water = 249.1 n/m^2 = 249.1 Pa
1 mm of mercury = 1 torr = 133 n/m^2 = 133 Pa
1 lbf/inch^2 = 6.895 kN/m^2 = 6.895 kPa
1kgf/cm^2 = 98.1kN/m^2 = 98.1 kPa

15

Important Expansions

ASAE : American Society of Agricultural Engineers

ASTM : American Society for Testing and Materials

BHP : Break Horse Power

BIS : Bureau of Indian Standards

BIS : Bureau of Indian Standards

DI : Direct Injection

I.C. engine : Internal Combustion engine

IHP : Indicated Horse Power

ISI : Indian Standards Institute

ISO : International Organization of Standards

KVIC : Khadi and Village Industries Commission

LVDT : Linear Variable Differential Transducer

MMD : Mean Mass Diameter

NMD : Number Mean Diameter

OECD : Organisation for Economic Co-operation and Development

OTEC : Ocean Thermal Energy Conversion

PTO : Power Take Off

RNAM : Regional Network for Agricultural Machinery

ROPS : Rollover Protection Structure

RTD : Resistance Temperature Detectors

SAE : Society of Automotive Engineers

SMAM : Submission of Agricultural Machinery

VMD : Volume Mean Diameter

References

Anonymous. (1983). RNAM Test Code & Procedures for Farm Machinery. Technical Series 12.

Bosoi ESO, Verniaev V, Smirnov and Sultan-Shakh EG. (1990). Theory,Construction and Calculations of Agricultural Machinery. Vol. I. Oxonian Press Pvt. Ltd. No.56.

e-Learning Portal on Agricultural Education https://ecourses.icar.gov.in

Indian Standard Codes for Agril. Implements. Published by ISI, New Delhi

H. Bernacki, J. haman, Cz. Kanafojske (1972). Agricultural machines, theory and construction Vol-I, USDA Publications, Warsaw, Poland.

Jain S.C and C.R.Rai (1985). Farm Tractors (Maintenance and Repairs). Tata Mc-Graw Hill Publication, New Delhi.

Jain R.K. (1994). Production Technology: A Text book for Engineering Students. Khanna Publishers, New Delhi.

Kalpak Jain, (2001). Manufacturing Process Engineering, Tata Mc Graw Hill Publication

Kepner R.A., Bainer R. and E.L.Barger (1987). Principles of Farm Machinery. CBS Publishers and Distributers, New Delhi.

Liljebahl J.B., Purnquist P.K. Smith D.V. and Makoto Hoki. (1996). Tractor and Their Power Unit. Society of Automotive Engineers, Inc., New Delhi.

Michael, A.M. and T.P. Ojha. (2016). Principles of Agricultural Engineering,Vol-I Jain Brothers Publishing House, New Delhi.

Metha M.L., Verma S.R., Mishra S.K. and Sharma V.K. (1995). Testing and Evaluation of Agricultural Machinery. National Agricultural Technology Information Centre, Ludhiana.

Nakra B.C. and Chaudhary K.K. (2004). Instrumentation Measurement and Analysis. Tata McGraw Hill.

O.P. Bindra H. Singh (1980). Pesticide Application Equipment, Oxford & IBH Publishing Co., New Delhi.

Rai G.D. (1994). Solar Energy Utilization. Khanna Publishers, New Delhi.

Sahay J. (1992). Elements of Agricultural Engineering,Agro Book Agency, New Chitragupta Nagar,Patna.

Shigley J.E. and Uicker J.J .(1980). Theory of Machinery and Mechanism. McGraw Hill.

Shrivastava A.C. (1991). Elements of Farm Machinery. Oxford and IBH Publishing Co. Pvt. Ltd., India.

Smith P.H. (1948). Farm Machinery and Equipments. Mc-Graw Hill Book Co. Inc., New York.

Sawhney A.K. (2008). Electrical & Electronics Measurement and Instrumentation. Dhanpat Rai & Sons

Sukhatme S.P. (1997). Solar Energy - Principles of Thermal Collection and Storage. 2nd Ed. Tata McGraw Hill.